An introduction
to numerical analysis

A title in the European Computer Science Series

Consulting Editor
Professor D. W. Barron
University of Southampton

An introduction to numerical analysis

C. W. Celia, M.Sc.
Principal Lecturer in Mathematics
Sir John Cass College, London

McGRAW-HILL · LONDON

New York · St. Louis · San Francisco · Sydney · Toronto
Mexico · Panama · Johannesburg

Published by

McGRAW-HILL Publishing Company Limited,
MAIDENHEAD · BERKSHIRE · ENGLAND

07 094115 7

PRINTED AND BOUND IN GREAT BRITAIN

Contents

Preface

This elementary introduction to numerical analysis is based on lectures given over the past ten years to undergraduates and graduates at Sir John Cass College. It is intended for mathematicians, scientists, engineers, and others who need a clear understanding of the basic methods in order to apply them in computation. It is particularly suitable for first-year undergraduates and for students taking numerical analysis as a subsidiary subject. There can be no doubt that familiarity with the general principles is an essential preliminary to programming numerical calculations.

The groundwork consists of the application of finite difference methods to the problem of interpolation, supplemented by the more direct approach by means of the Lagrangian interpolation polynomial. This leads to numerical integration including Gaussian quadrature, with estimates of the truncation error. The treatment of the solution of ordinary differential equations includes a discussion of stability. For use in sub-routines, approximations to known functions are obtained in the form of series of orthogonal polynomials. Examples are given of the application of Chebyshev polynomials to the solution of a class of differential equations. Stress is laid on the importance of iterative methods in general, and in particular in the solution of non-linear equations. For the solution of a system of linear algebraic equations, both direct and iterative methods are considered, and are related to the use of matrices.

There are many worked examples explained in detail and each chapter contains a number of examples for the reader himself to work. Only by practice and experience can skill in the art of computing be acquired.

C. W. CELIA

1

Introduction

In numerical analysis our purpose is to devise methods for obtaining numerical answers to mathematical problems. For example we may require answers to questions such as these:

(*a*) Find correct to five significant figures the roots of the equation

$$x^2 - 20x + 3 = 0$$

(*b*) Find the value of y for $x = 0(0{\cdot}1)1{\cdot}0$, (i.e., for values of x from 0 to 1 by steps of $0{\cdot}1$), correct to four significant figures, given that

$$dy/dx = x^2 + y^2, \quad y = 0 \text{ when } x = 0$$

(*c*) Solve the system of simultaneous equations

$$0{\cdot}41x_1 + 0{\cdot}38x_2 + 0{\cdot}62x_3 = 0{\cdot}48$$
$$0{\cdot}38x_1 - 0{\cdot}81x_2 - 0{\cdot}46x_3 = 0{\cdot}21$$
$$0{\cdot}62x_1 - 0{\cdot}46x_2 + 0{\cdot}78x_3 = 0{\cdot}39$$

(*d*) If $f(0) = 2{\cdot}34$, $f(1) = 2{\cdot}59$, $f(3) = 2{\cdot}91$ estimate the value of $f(2)$.

(*e*) Evaluate, for $x = 0(0{\cdot}1)2$, $\int_0^x \sin(x^2)\, dx$ correct to six decimal places.

In the first example, we find, by using tables giving four decimal places, the result $x = 10 \pm \sqrt{97} = 19{\cdot}8489$ or $0{\cdot}1511$. This gives the larger root to be $19{\cdot}849$ to five significant figures, but it fails to give the smaller root to the same accuracy. To obtain this we need to change our method. Since the product of the roots is 3, the smaller root is equal to $3/19{\cdot}8489$, which gives $0{\cdot}15114$ as the required value. In example (*b*) the analytic solution is not easy to find, and even when found, it is difficult to evaluate. A direct numerical solution is both quicker and simpler. In example (c) the coefficients of the unknowns have been rounded-off to two decimal places. If the coefficients were exact, there would be a single exact solution, but since the coefficients are rounded-off, no exact solution exists. The reliability of our solution will depend on what effect small changes in the coefficients have upon the solution. In example (d) we have to assume that $f(x)$ is a quadratic function of x in order to estimate $f(2)$, and the reliability of our estimated value depends on the justification of this assumption. In the last example, we are asked to tabulate the values of an integral which cannot be evaluated analytically.

The results of our calculations will be given in decimal form to a required number of places, so that they are approximations to the exact answers. We define the error in an approximation to be

$$\text{(the exact answer)} - \text{(the calculated answer)}$$

It is essential that we have some idea of its magnitude, so that we can judge the reliability of our result. The size of the error will depend on the accuracy of the data, on the methods we choose to employ, and on the detection of all blunders. We now consider these three factors in turn.

(a) Accuracy of the data

In general, the values given are stated in decimal form, rounded-off to a given number of places. The constant e for example may be quoted as $2 \cdot 7183$ to four decimal places, or as $2 \cdot 72$ to two decimal places. If t is the number of decimal places, the round-off error lies between $(-\frac{1}{2})10^{-t}$ and $(+\frac{1}{2})10^{-t}$. Note that if the only figure to be rounded-off is a 5, then we end the rounded number with an even figure, e.g.,

$$2 \cdot 635 \text{ and } 2 \cdot 645 \text{ become } 2 \cdot 64$$
$$2 \cdot 655 \text{ and } 2 \cdot 665 \text{ become } 2 \cdot 66$$

This round-off in the data introduces a certain inexactness at the outset, and care must be taken not to quote results to a degree of accuracy that is not justifiable. It is not usually possible to arrive at an answer which has more significant figures than the data, though this is not impossible. If we are given $x = 9 \cdot 2$, rounded to one decimal place, and we calculate $x^{0 \cdot 2}$, we can be certain that the result lies between $(9 \cdot 25)^{0 \cdot 2}$ and $(9 \cdot 15)^{0 \cdot 2}$, so that our answer is $1 \cdot 56$, correct to two decimal places.

(b) The choice of methods

The methods we use must be practical and must be completed in a finite number of steps. They will almost inevitably introduce some error, and it requires skill to keep this error to a minimum. To illustrate this point we evaluate

$$x/\{\sqrt{(x + 1)} + \sqrt{x}\}$$

when x is exactly 6, working to three decimal places.

Method (i): $\sqrt{7} + \sqrt{6} = 2\cdot646 + 2\cdot449 = 5\cdot095$
$6 \div 5\cdot095 = 1\cdot178$
Method (ii): Rationalizing gives
$$6(\sqrt{7} - \sqrt{6}) = 6 \times 0\cdot197 = 1\cdot182$$

There is no mistake here, yet the answers differ. If we round-off to two decimal places, both methods give the answer $1\cdot18$. (If we repeat the calculation with $x = 7$, we arrive at the answers $1\cdot279$ and $1\cdot274$, which will not agree when rounded-off.) The weakness of method (ii) lies in the fact that the difference $(\sqrt{7} - \sqrt{6})$ contains only three significant figures. In the number $0\cdot197$ there is a round-off error of over $0\cdot3\%$. We can place no reliance in the final figure of the answer $1\cdot182$ and should give the answer to three significant figures only. It is very important to avoid wherever possible the loss of accuracy caused by taking the difference between two numbers which are approximately equal.

It is worth noting that multiplication usually leads to round-off error. Consider the evaluation of

$$(x - 0\cdot5)(x + 1\cdot7) + (x + 0\cdot2)(x - 0\cdot6)$$

when $x = 4\cdot2$, rounded to one decimal place.

Method (i): $3\cdot7 \times 5\cdot9 = 21\cdot8$, rounded-off to one decimal place
$4\cdot4 \times 3\cdot6 = 15\cdot8$, rounded-off to one decimal place
$21\cdot8 + 15\cdot8 = 37\cdot6$
Method (ii): $(3\cdot7 \times 5\cdot9) + (4\cdot4 \times 3\cdot6) = 21\cdot83 + 15\cdot84 = 37\cdot67$ which gives $37\cdot7$ when rounded-off.

In method (ii) we have a simple example of the 'accumulation of products'. It is more accurate to round-off the final result, and not the intermediate results.

A second example of a choice of method is provided by the evaluation of a polynomial. Suppose that we require the value when $x = k$ of the polynomial

$$a_0 + a_1x + a_2x^2 + a_3x^3 + a_4x^4$$

Instead of calculating each term separately and then adding, we find it better to calculate in turn b_4, b_3, b_2, b_1, where

$$b_4 = a_4k$$
$$b_3 = (b_4 + a_3)k$$
$$b_2 = (b_3 + a_2)k$$
$$b_1 = (b_2 + a_1)k + a_0$$

With a desk-machine, this has the advantage that no intermediate result has to be recorded.

We must also consider the truncation error that arises in a method using formulae that are not exact. For instance, if in a computer program the value of sin (0·1) is required correct to six decimal places, it could be found by means of the relation

$$\sin x = x - x^3/6 + x^5/120$$

This formula is not exact since the series has been truncated, i.e., cut off, after the term in x^5. It will however give the required value of sin (0·1), since the terms neglected will not affect the sixth decimal place when $x = 0·1$. If we need the value of sin x at $x = 1$ and if we use only the same three terms as before, our result will not be correct to six decimal places. There will be a truncation error equal approximately to $-1/7!$, since the first neglected term in the series for sin x is $-x^7/7!$.

Another example of a truncation error is given by Simpson's rule for integration

$$\int_0^{2h} y \, dx = \tfrac{1}{3}h\{y(0) + 4y(h) + y(2h)\}$$

If y is a polynomial in x of degree not higher than three, this is an exact relation. Otherwise there is a truncation error proportional to h^5.

Many problems arise that cannot be solved in a finite number of steps. For these we need a method in which the result can be obtained to any specified degree of accuracy by increasing the number of steps in the calculation. In view of the great importance of the iterative type of method we consider the calculation of the cube root of a positive number N correct to five decimal places, by means of a sequence $x_1, x_2, x_3, \ldots$ defined by the relation

$$x_{n+1} = (2x_n^3 + N)/3x_n^2, \quad \text{with } x_1 = N$$

If we put $N = k^3$, we find

$$
\begin{aligned}
x_{n+1} - k &= (2x_n^3 - 3kx_n^2 + k^3)/3x_n^2 \\
&= (x_n - k)(2x_n^2 - kx_n - k^2)/3x_n^2 \\
&= (x_n - k)^2(2x_n + k)/3x_n^2
\end{aligned}
$$

This shows that x_{n+1} is greater than k for all values of n, and that

$$(x_{n+1} - k) \leqslant \tfrac{2}{3}(x_n - k) \leqslant (\tfrac{2}{3})^2(x_{n-1} - k) \leqslant (\tfrac{2}{3})^n(x_1 - k)$$

We can therefore make the difference between x_{n+1} and k as small as we please by increasing n sufficiently. On this basis we can formulate an algorithm for the calculation of the cube root, i.e., we can set out step by step the operations that will lead to the result. This algorithm can be displayed in the form of a flow diagram (Fig. 1.1).

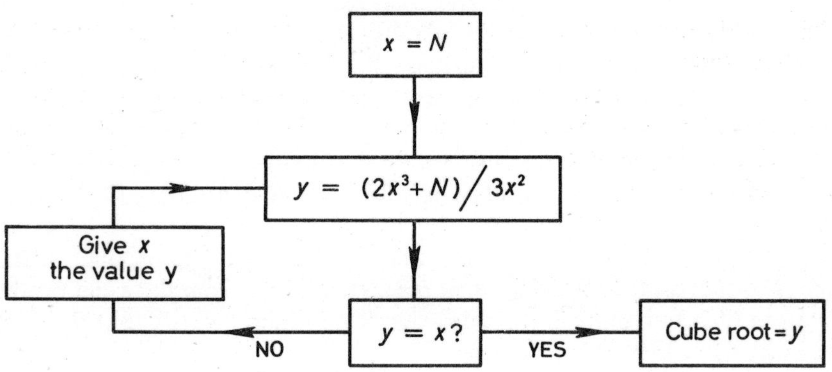

Fig. 1.1 An algorithm in the form of a flow diagram

The steps are as follows:

  (i) Take x_1 equal to N.
 (ii) With $x - x_1$, calculate y to six decimal places. This gives the value of x_2.
(iii) Test whether $x_2 = x_1$.
(iv) If so, the result is given by x_2 rounded-off to five decimal places.
 (v) If not, with $x = x_2$ recalculate y. This gives x_3.
(vi) Test whether $x_3 = x_2$.

This iteration continues until $x_{n+1} = x_n$ to six decimal places. With $N = 3$, we obtain the sequence

$$x_1 = 3$$
$$x_2 = 2{\cdot}111\ 111$$
$$x_3 = 1{\cdot}631\ 784$$
$$x_4 = 1{\cdot}463\ 412$$
$$x_5 = 1{\cdot}442\ 554$$
$$x_6 = 1{\cdot}442\ 250$$
$$x_7 = 1{\cdot}442\ 250$$

Thus to five decimal places $3^{\frac{1}{3}} = 1{\cdot}44225$.

(*c*) The detection of blunders

We use the word 'error' to cover all inaccuracies introduced in the data or arising from the method. We use the word 'mistake' to denote a blunder or a slip on the part of the machine or of its operator. No one is infallible, least of all the person who imagines himself to be so. It is essential to check each step in a calculation. If a mistake is overlooked, all subsequent work may be wasted. Wherever possible the checks should be built into the method, and they should never rely on the identical repetition of the calculation. To give a result without checking is to ask for trouble.

EXAMPLES

(1) Find the range of positive values of x for which the truncated series $(1 + x + x^2/2! + x^3/3! + x^4/4!)$ will give the values of exp x correct to four decimal places.

(2) Find the maximum error in the calculated value of ab/c, given that the values of a, b, and c rounded-off to two decimal places are $1·84$, $1·80$, and $1·15$ respectively.

(3) If x and y are positive and less than 1, show that the product of their values rounded-off to three decimal places can differ from the exact product by more than five units in the third significant figure. Verify that this is the case when $x = 0·9455$, $y = 0·1035$.

(4) Show that $(b^2 - ac)/(a - 2b + c)$ and $[(a - b)^2/(a - 2b + c) - a]$ are identically equal. Compare the ease with which these expressions can be evaluated

    (*a*) when $a = 0·46, \quad b = 0·48, \quad c = 0·49$
    (*b*) when $a = 0·4604, b = 0·4802, c = 0·4901$

(5) Show that the sequence defined by $x_1 = k^2 \ (k > 0)$, $x_{n+1} = \frac{1}{2}(x_n + k^2/x_n)$, will converge to k. Set out a flow diagram for the calculation of the square root of a positive number. Calculate $\sqrt{5}$ to five decimal places.

2
Finite differences

Suppose that the distances travelled by a racing car in five successive seconds are measured and found to be 90 ft, 98 ft, 106 ft, 114 ft, 122 ft. Then it is clear that the speed is increasing at a constant rate, because the difference between consecutive values is always 8 ft. In another case let the distances be 90 ft, 99 ft, 110 ft, 123 ft, 138 ft. Here the difference is not constant. We form two difference tables, one for each case.

(a)			(b)		
90			90		
	8			9	
98			99		2
	8			11	
106			110		2
	8			13	
114			123		2
	8			15	
122			138		

The distances are entered in the first column and the values in the first difference column are found by subtracting each distance from the one beneath it. In (b) we obtain the second difference column from the first difference column, and in this example the values in the second difference column are all equal. There is a similarity here between differencing and differentiating. If dy/dx is constant we deduce that y is increasing at a constant rate so that there is a relation between x and y of the form $y = ax + b$. In the same way we see that the distance s is related to the time t by the equation $s = 8t + 90$. When d^2y/dx^2 is constant we deduce the relation

$$y = ax^2 + bx + c$$

Similarly the values in (b) satisfy the equation

$$s = t^2 + 8t + 90$$

To see why this is so, consider the values of x^n for $x = 0, 1, 2, \ldots$. The first difference column will begin with the entries 1^n, $(2^n - 1^n)$, $(3^n - 2^n)$, and we can write the general term as $(x + 1)^n - x^n$. Since this is a polynomial of degree $n - 1$, the effect of the differencing is to reduce the degree from n to $n - 1$. If we had begun with a polynomial of degree n instead of the single term x^n, the same result would have been reached. If we form the second difference column, again the degree is reduced by one. We deduce that the nth-order differences of a polynomial of degree n are all equal. In particular, the second-order differences of a quadratic expression are all equal.

We have so far assumed that the values of x and y are integers, so that all our values are exact. To see what happens when the values of y are rounded-off we form a difference table for x^2 for $x = 1, 1{\cdot}02, 1{\cdot}04, \ldots, 1{\cdot}1$. Notice that in the columns of differences we do not use the decimal point.

(a) Exact Values   1·0000

	404	
1·0404		8
	412	
1·0816		8
	420	
1·1236		8
	428	
1·1664		8
	436	
1·2100		

(b) Values rounded to three decimal places

1·000		
	40	
1·040		2
	42	
1·082		0
	42	
1·124		0
	42	
1·166		2
	44	
1·210		

When the values are rounded-off, we no longer obtain the column of constant second differences, and if we now form the column of third differences, the values we obtain are quite useless. They arise solely from the round-off errors in the values of x^2. By analogy with telephone circuitry, we speak of such values as 'noise'. There is no point in extending a difference table to extra columns once this 'noise level' is reached.

The worst possible case arises when the values of y, rounded-off to t decimal places, have the maximum error, $\frac{1}{2} \times 10^{-t}$, with consecutive errors opposite in sign. Then each of the first differences has an error 1×10^{-t}, each of the second differences an error 2×10^{-t}, each of the third differences an error 4×10^{-t} and so on. This gives a guide to the size of irregularities to expect.

Forward differences

We denote by h the distance between consecutive values of x, so that if the first value is x_0, then $x_1 = x_0 + h$, $x_2 = x_0 + 2h$, $x_n = x_0 + nh$. The corresponding values of y we denote by y_0, y_1, y_2, y_n so that $y_r = y(x_0 + rh)$. The distance h is known as the step-length.

The first forward differences are given by

$$\Delta y_0 = y_1 - y_0, \qquad \Delta y_1 = y_2 - y_1, \qquad \Delta y_n = y_{n+1} - y_n$$

The second differences are given by

$$\Delta^2 y_0 = \Delta y_1 \quad - \Delta y_0 = y_2 \quad - 2y_1 \quad + y_0$$
$$\Delta^2 y_1 = \Delta y_2 \quad - \Delta y_1 = y_3 \quad - 2y_2 \quad + y_1$$
$$\Delta^2 y_n = \Delta y_{n+1} - \Delta y_n = y_{n+2} - 2y_{n+1} + y_n$$

In the same way

$$\Delta^m y_0 = \Delta^{m-1} y_1 - \Delta^{m-1} y_0$$

With this notation the difference table takes the form:

$$
\begin{array}{ccccccccc}
y_0 \\
& \Delta y_0 \\
y_1 & & \Delta^2 y_0 \\
& \Delta y_1 & & \Delta^3 y_0 \\
y_2 & & \Delta^2 y_1 & & \Delta^4 y_0 \\
& \Delta y_2 & & \Delta^3 y_1 \\
y_3 & & \Delta^2 y_2 \\
& \Delta y_3 \\
y_4
\end{array}
$$

Backward differences

We put

$$\nabla y_1 = y_1 - y_0, \qquad \nabla y_2 = y_2 - y_1, \qquad \nabla y_n = y_n - y_{n-1}$$
$$\nabla^2 y_2 = \nabla y_2 - \nabla y_1 = y_2 - 2y_1 + y_0$$
$$\nabla^2 y_n = \nabla y_n - \nabla y_{n-1} = y_n - 2y_{n-1} + y_{n-2}$$
$$\nabla^m y_1 = \nabla^{m-1} y_1 - \nabla^{m-1} y_0$$

With this notation the difference table reads

$$
\begin{array}{ccccc}
y_0 & & & & \\
& \nabla y_1 & & & \\
y_1 & & \nabla^2 y_2 & & \\
& \nabla y_2 & & \nabla^3 y_3 & \\
y_2 & & \nabla^2 y_3 & & \nabla^4 y_4 \\
& \nabla y_3 & & \nabla^3 y_4 & \\
y_3 & & \nabla^2 y_4 & & \\
& \nabla y_4 & & & \\
y_4 & & & &
\end{array}
$$

When the numerical values of y are put in, and the differences are formed, the forward difference table and the backward difference table are identical. We have two names for each number; for example, $\Delta^3 y_1$ and $\nabla^3 y_4$ are the same number. The purpose of this will be seen later in the applications of differences.

Operators

We speak of Δ and ∇ as forward and backward difference operators. From our definitions we have

$$\Delta(\Delta y_n) = \Delta^2 y_n, \qquad \nabla(\nabla y_n) = \nabla^2 y_n$$

Also

$$\Delta(\nabla y_n) = \Delta(y_n - y_{n-1}) = \Delta y_n - \Delta y_{n-1} = \nabla(\Delta y_n)$$

so that the order in which Δ and ∇ are applied makes no difference.

A new operator, E, known as the shift operator, is defined by

$$Ey_0 = y_1, \qquad E^{-1}y_1 = y_0, \qquad Ey_n = y_{n+1}$$

It has the effect of increasing the value of x by h, the step-length. It is simple to show that $\Delta Ey_0 = E\Delta y_0$, and $\nabla Ey_0 = E\nabla y_0$. Now $\Delta y_0 = y_1 - y_0 = Ey_0 - y_0 = (E - 1)y_0$ so that Δ and $(E - 1)$ are equivalent operators, and $E = 1 + \Delta$.

Also $\nabla y_1 = y_1 - y_0 = (E - 1)y_0$, so that $\nabla E = E - 1$. We can express this in the form $E = (1 - \nabla)^{-1}$.

If we now operate on a polynomial y of degree m (with integral values), this gives

$$
\begin{aligned}
Ey_n &= (1 - \nabla)^{-1} y_n \\
&= (1 + \nabla + \nabla^2 + \cdots) y_n
\end{aligned}
$$

i.e.,
$$y_{n+1} = y_n + \nabla y_n + \nabla^2 y_n + \cdots + \nabla^m y_n$$

since all higher order differences will be zero.

We can apply this usefully to a difference table for x^3 starting with the values for $x = 1{\cdot}0, 1{\cdot}1, 1{\cdot}2, 1{\cdot}3$.

$$
\begin{array}{llll}
y_0 & 1{\cdot}000 & & \\
& & 331 & \\
y_1 & 1{\cdot}331 & & 66 \\
& & 397 & & 6 \\
y_2 & 1{\cdot}728 & & 72 \\
& & 469 & \\
y_3 & 2{\cdot}197 & &
\end{array}
$$

We have
$$
\begin{aligned}
y_4 &= y_3 + \nabla y_3 + \nabla^2 y_3 + \nabla^3 y_3 \\
&= 2{\cdot}197 + 0{\cdot}469 + 0{\cdot}072 + 0{\cdot}006 = 2{\cdot}744
\end{aligned}
$$

Alternatively, since the third differences are constant, we can put in the values $\nabla^3 y_4 = 6$, $\nabla^2 y_4 = 78$, $\nabla y_4 = 547$, so that $y_4 = 2{\cdot}197 + 0{\cdot}547 = 2{\cdot}744$.

The advantage of extending a table in this way is that addition on a desk machine takes less time than multiplication. If the table above is continued to the value $x = 2$, the corresponding value of y is known to be 8, so that we have a check on the intermediate calculations. We naturally use backward differences in this building-up process and this is a typical use for them. It will be seen later that in solving differential equations we use backward differences to extend a difference table downwards even though we may not have the advantage of a column of constant differences as above.

Central differences

The central difference operator δ is defined by

$$
\delta = E^{\frac{1}{2}} - E^{-\frac{1}{2}}
$$

so that $\delta y_{\frac{1}{2}} = y_1 - y_0$. Since $\delta^2 = E - 2 + E^{-1}$ we have

$$
\begin{aligned}
\delta^2 y_0 &= y_1 - 2y_0 + y_{-1} = \Delta^2 y_{-1} = \nabla^2 y_1 \\
\delta^2 y_1 &= y_2 - 2y_1 + y_0 \; = \Delta^2 y_0 \; = \nabla^2 y_2
\end{aligned}
$$

With central difference notation the difference table reads as follows:

$$
\begin{array}{llll}
y_0 & & & \\
& \delta y_{\frac{1}{2}} & & \\
y_1 & & \delta^2 y_1 & \\
& \delta y_{\frac{3}{2}} & & \delta^3 y_{\frac{3}{2}} \\
y_2 & & \delta^2 y_2 & \\
& \delta y_{\frac{5}{2}} & & \\
y_3 & & &
\end{array}
$$

Notice that in the first differences the suffix contains the fraction $\frac{1}{2}$, while in the second differences the suffix is always an integer. Central differences with the same suffix lie on the same horizontal line in the table. Where the values of the required differences are available, a central difference formula is in general to be preferred to a forward or a backward difference formula. At the head of a table we are forced to use the forward differences, while at the foot only backward differences are available. In the body of a table we can employ central difference formulae. The reason for their superiority will be seen later.

So far we have always denoted by y_0 the first value of y in the table, but often it is convenient to have y_0 in the body of the table. This is illustrated by the following table.

$$
\begin{array}{ccccccc}
y_{-2} \\
& \delta y_{-\frac{3}{2}} \\
y_{-1} & & \delta^2 y_{-1} \\
& \delta y_{-\frac{1}{2}} & & \delta^3 y_{-\frac{1}{2}} \\
y_0 & & \delta^2 y_0 & & \delta^4 y_0 \\
& \delta y_{\frac{1}{2}} & & \delta^3 y_{\frac{1}{2}} \\
y_1 & & \delta^2 y_1 \\
& \delta y_{\frac{3}{2}} \\
y_2
\end{array}
$$

In order to cope with expressions such as δy_0 and $\delta^2 y_{\frac{1}{2}}$, which do not correspond to entries in the difference table, we now introduce the averaging operator μ. This is defined by the relation

$$\mu y_0 = \tfrac{1}{2}(y_{\frac{1}{2}} + y_{-\frac{1}{2}})$$

so that the operators μ and $\frac{1}{2}(E^{\frac{1}{2}} + E^{-\frac{1}{2}})$ are equivalent. It follows that

$$\mu^2 = \tfrac{1}{4}(E + 2 + E^{-1}) = 1 + \tfrac{1}{4}\delta^2$$

Since
$$\mu\delta = \tfrac{1}{2}(E^{\frac{1}{2}} + E^{-\frac{1}{2}})(E^{\frac{1}{2}} - E^{-\frac{1}{2}}) = \tfrac{1}{2}(E - E^{-1})$$

and
$$\delta\mu = \tfrac{1}{2}(E^{\frac{1}{2}} - E^{-\frac{1}{2}})(E^{\frac{1}{2}} + E^{-\frac{1}{2}}) = \tfrac{1}{2}(E - E^{-1})$$

we have
$$\mu\delta = \delta\mu$$

We can now write

$$\mu\delta y_0 = \delta\mu y_0 = \tfrac{1}{2}(\delta y_{\frac{1}{2}} + \delta y_{-\frac{1}{2}}) = \tfrac{1}{2}(y_1 - y_{-1})$$

The δ^2 process

If $x_n = a + br^n$, and the sequence $\{x_n\}$ converges, the limit a of the sequence can be found by means of a method known as the δ^2 process. We have

$$\delta^2 x_n = x_{n+1} - 2x_n + x_{n-1} = b(1 - r)^2 r^{n-1}$$

and
$$(x_{n+1} - x_n)^2 = b^2(1 - r)^2 r^{2n}$$

Hence
$$x_{n+1} - (x_{n+1} - x_n)^2/\delta^2 x_n = x_{n+1} - br^{n+1} = a$$

Thus from three consecutive values x_{n-1}, x_n, x_{n+1} we can obtain the value of the limit of the sequence. Notice that the accuracy of the result is limited by the number of significant figures in the second difference. The δ^2 process is widely used for accelerating the convergence of iterative processes in which the relation $x_n = a + br^n$ is only approximately true.

EXAMPLES

(1) Five consecutive members of a sequence are $x_1 = 3\cdot33718$, $x_2 = 3\cdot20993$, $x_3 = 3\cdot09479$, $x_4 = 2\cdot99061$, $x_5 = 2\cdot89634$. Find the limit of the sequence by applying the δ^2 process to x_1, x_2, x_3, to x_2, x_3, x_4, and to x_3, x_4, x_5.

(2) Form difference tables for x^4 and x^5 for integral values of x from -4 to $+4$.

(3) Form a difference table for the function $\cosh x$, starting with the values $\cosh 0 = 1$, $\cosh (\pm0\cdot01) = 1\cdot00005$, on the assumption that the second differences are constant. Extend the table to give the value of $\cosh (1\cdot1)$ and check the result from tables.

(4) If s_n denotes the sum of the first n terms of the series

$$1 - \tfrac{1}{3} + \tfrac{1}{5} - \tfrac{1}{7} + \tfrac{1}{9} - \tfrac{1}{11} + \cdots$$

apply the δ^2 process to s_6, s_7, and s_8. Compare the result with $\tfrac{1}{4}\pi$.

(5) Form a difference table for $y = x^4$ starting with the values of y at $x = 0$, $\pm0\cdot1$, $\pm0\cdot2$. Extend it to $x = 1$.

(6) Form difference tables for $y = 2^x$ and $y = \sin\left(\tfrac{1}{2}\pi x\right)$ with $x = 0, 1, 2, \ldots$.

The step-length h

In choosing a suitable value for h we have to remember that if h is too small the amount of work required is excessive, and that if h is too large the higher order differences will not be negligible. In many formulae in common use, the differences of order higher than the fourth are neglected, and h must be chosen to satisfy this requirement. The effect of the size of h on the differences is easy to see. If we double h, then in place of the first forward difference $y_1 - y_0$, we have $y_2 - y_0$, so that the first forward difference will be doubled approximately. Suppose that when h is replaced by $\tfrac{1}{2}h$, the new shift operator is denoted by E_1 and the new forward difference operator is denoted by Δ_1. We have

$$E_1 = E^{\frac{1}{2}} = (1 + \Delta)^{\frac{1}{2}} = 1 + \tfrac{1}{2}\Delta - \tfrac{1}{8}\Delta^2 + \cdots$$

Hence

$$\Delta_1 = \tfrac{1}{2}\Delta - \tfrac{1}{8}\Delta^2 + \cdots, \qquad \Delta_1^2 = \tfrac{1}{4}\Delta^2 - \cdots, \qquad \Delta_1^3 = \tfrac{1}{8}\Delta^3 - \cdots$$

so that if the step-length is halved, the approximate result is to divide the first, second, and third differences by 2, 4, 8 respectively.

We have already seen that if the values of y are exact values of a polynomial then the difference table terminates. When $y = x^n$ and $h = 1$, $\Delta^n y = n!$ and $\Delta^{n+1} y = 0$. If y is a polynomial of degree n, but the values of y are rounded-off, then we fail to obtain $\Delta^{n+1} y = 0$. Instead, the differences of order $(n + 1)$ consist purely of 'noise', and higher order differences can be large. If y is not a polynomial in x, we cannot expect to obtain a column in the difference table consisting only of zeros. Even if we take exact values of y, we cannot take it for granted that the magnitude of the differences will tend to decrease as the order rises. Clearly the value of h will have some influence.

Consider the exponential function $y = e^x$. We have

$$\Delta y = e^{x+h} - e^x = e^x(e^h - 1)$$
$$\Delta^2 y = (e^{x+2h} - e^{x+h}) - (e^{x+h} - e^x) = e^x(e^h - 1)^2, \quad \text{etc.}$$

Then $\Delta^n y = e^x(e^h - 1)^n$, and so it follows that $\Delta^n y$ will tend to zero only if $(e^h - 1)$ is less than 1, and for this we need $h < 0.6931$. If h is larger than this value, the table will diverge. When h is small, $(e^h - 1)$ equals h approximately, so that $\Delta y = hy$, $\Delta^n y = h^n y$.

As a second example, consider the function $\sin x$.

$$\Delta (\sin x) = \sin (x + h) - \sin (x)$$
$$= 2 \cos (x + \tfrac{1}{2}h) \sin (\tfrac{1}{2}h)$$
$$\Delta^2 (\sin x) = 2 \sin (\tfrac{1}{2}h) \{\cos (x + \tfrac{3}{2}h) - \cos (x + \tfrac{1}{2}h)\}$$
$$= -4 \sin^2 (\tfrac{1}{2}h) \sin (x + h)$$

Similarly $\Delta^2 (\cos x) = -4 \sin^2 (\tfrac{1}{2}h) \cos (x + h)$

Here we need $2 \sin (\tfrac{1}{2}h)$ less than 1, which will be true if $h < 1$ since $\sin 0.5 < \sin 30° = \tfrac{1}{2}$. We must choose h so that $\Delta^n y$ will tend to zero as n increases (when exact values of y are used). In practice, of course, the effect of round-off will always result in eventual divergence.

Detection of mistakes

We are so accustomed to accept without question the values given in a printed table, that it comes as a shock when we detect a mistake. Such blunders often arise when a table is copied by hand, or when the values are being rounded-off to the required number of decimal places. The commonest mistake is the transposition of two adjacent digits.

The table below on the left gives values of $\cos x$ for $x = 0.73(0.01)0.79$. It contains a slip that remains uncorrected in the latest edition of a well-known book of tables.

When the difference table is formed, we expect the value of $\Delta^2 \cos x$ to be $-10^{-4} \cos x$, but find that the second differences are very irregular. The maximum error from round-off in the second differences is 2 units, so that

x	$\cos x$			$\cos x$		
0·73	0·74517			0·74517		
		−670			−670	
0·74	0·73847		−8	0·73847		−8
		−678			−678	
0·75	0·73169		−3	0·73169		−7
		−681			−685	
0·76	0·72488		−16	0·72484		−8
		−697			−693	
0·77	0·71791		−3	0·71791		−7
		−700			−700	
0·78	0·71091		−6	0·71091		−6
		−706			−706	
0·79	0·70385			0·70385		

some other explanation must be found. The value of cos x for $x = 0.76$ should be 0·72484, and from this value we obtain the correct differences shown on the right-hand side. The change of 4 units in the value of cos x throws out two of the first differences by 4 units and it affects three of the second differences by 4, 8, and 4 units respectively. The way in which the effect fans out can be seen from the following difference table, showing a single change e in one value of y. Note especially the binomial coefficients, and the fact that the sum of the entries in any column of differences is zero.

```
0
    0
0           0
    0            0
0           0           e
    0           e
0          e          −4e
    e          −3e
e         −2e          6e
   −e           3e
0           e          −4e
    0           −e
0           0           e
    0            0
0           0
    0
0
```

In the difference table for cos x we can argue that the average value of the second differences will be unchanged and that the disturbance will show itself in three consecutive terms in the form e, $-2e$, e. The average value is

-7 units to the nearest whole number. The disturbance is then seen to be $+4, -7, +4$ centred on the line for $x = 0.76$. It follows that $e = +4$ units and from this we can make the correction and check it by differencing.

In the following table for $y = \exp x$, $x = 0(0.1)1$, a mistake has been made in one value.

x	y				
0	1·00000				
		10517			
0·1	1·10517		1106		
		11623		117	
0·2	1·22140		1223		10
		12846		127	
0·3	1·34986		1350		17
		14196		144	
0·4	1·49182		1494		21
		15690		165	
0·5	1·64872		1659		−19
		17349		146	
0·6	1·82221		1805		74
		19154		220	
0·7	2·01375		2025		−18
		21179		202	
0·8	2·22554		2227		33
		23406		235	
0·9	2·45960		2462		
		25868			
1·0	2·71828				

The fourth differences indicate a mistake centred on the line for $x = 0.6$. We can argue that since the average

$$(21 - 19 + 74 - 18 + 33)/5$$

is about 18, the change in $\Delta^4 y(0.6)$ is $(74 - 18)$ units. This should equal $6e$, as the relevant binomial coefficients are $1, -4, 6, -4, 1$. Hence e is $(74 - 18)/6$, i.e., 9 units. The corrected value of y at $x = 0.6$ is 1.82212 and the column of fourth differences will then read 10, 17, 12, 17, 20, 18, 24.

This set of values is not particularly smooth, but it must be remembered that the error due to round-off can be as large as 8 units in the fourth differences. It is worth noticing that the table above illustrates the relation $\Delta^n(e^x) = h^n e^x$ which we had earlier; for example, the second differences are approximately one hundredth of the corresponding values of y. In the solution of a differential equation on a desk-machine by a step-by-step method, the difference table plays an important part and it enables us to detect mistakes such as the foregoing. A sudden irregularity in a column of differences must always be investigated.

Fitting a polynomial to given values

Two points on a graph are enough to determine a straight line, three points are enough for a quadratic expression and four points are enough to determine a cubic. In general, $(n + 1)$ points are sufficient to enable us to find the $(n + 1)$ coefficients in a polynomial of degree n. When the values of x are equally spaced, we can employ forward differences for this purpose.

Suppose that we require a curve with equation

$$y = ax^3 + bx^2 + cx + d$$

which is satisfied by the pairs of values $(0, -1)$, $(1, -3)$, $(2, 1)$, $(3, 17)$. We form the difference table for these values:

x	y	Δy	$\Delta^2 y$	$\Delta^3 y$
0	-1			
		-2		
1	-3		6	
		4		6
2	1		12	
		16		
3	17			

Then

$$y_n = E^n y_0 = (1 + \Delta)^n y_0$$
$$= y_0 + n\Delta y_0 + \tfrac{1}{2}n(n - 1)\Delta^2 y_0 + \tfrac{1}{6}n(n - 1)(n - 2)\Delta^3 y_0$$

There are no more terms since $\Delta^4 y$ is zero when y is a cubic polynomial. By substituting $\Delta y_0 = -2$, $\Delta^2 y_0 = 6$, $\Delta^3 y_0 = 6$ we obtain

$$y_n = n^3 - 3n - 1$$

It is easily verified that the values $n = 0, 1, 2, 3$ give y the required values $-1, -3, 1, 17$. We now replace n by x (since $x = nh$ and $h = 1$) and write $y = x^3 - 3x - 1$. This polynomial fits the given values.

EXAMPLES

(1) Show that $\delta = \Delta(1 + \Delta)^{-\frac{1}{2}} = \nabla(1 - \nabla)^{-\frac{1}{2}}$.

(2) Prove that

$$\Delta^3 y_0 = y_3 - 3y_2 + 3y_1 - y_0$$
$$\nabla^3 y_0 = y_0 - 3y_{-1} + 3y_{-2} - y_{-3}$$
$$\delta^2 y_0 = y_1 - 2y_0 + y_{-1}$$
$$\delta^4 y_0 = y_2 - 4y_1 + 6y_0 - 4y_{-1} + y_{-2}$$
$$\mu\delta y_0 = \tfrac{1}{2}(y_1 - y_{-1})$$
$$\mu\delta^3 y_0 = \tfrac{1}{2}(y_2 - 2y_1 + 2y_{-1} - y_{-2})$$

(3) By expressing each in terms of E, show that

$$\Delta(1 + \tfrac{1}{2}\Delta)/(1 + \Delta), \quad \nabla(1 - \tfrac{1}{2}\nabla)/(1 - \nabla)$$

and $\mu\delta$ are equivalent. Obtain the expansion of $\mu\delta$ as

(*a*) a series of powers of Δ, (*b*) a series of powers of ∇.

(4) Correct the mistakes in the following tables of values:

x	$\exp x$	$\cosh x$	$\sinh x$
1·0	2·7183	1·5431	1·1752
1·1	3·0042	1·6685	1·3356
1·2	3·3201	1·8107	1·5059
1·3	3·6693	1·9790	1·6984
1·4	4·0525	2·1509	1·9043
1·5	4·4817	2·3524	2·1293
1·6	4·9530	2·5775	2·3756

(5) Find the cubic polynomial $y(x)$ which takes the following values:

$$y(0) = 0, \qquad y(1) = -4, \qquad y(2) = -8, \qquad y(3) = 0$$

(6) Find the quadratic in x that takes the same values as the function $\cos(\pi x/6)$ at the points $x = 2, 3, 4$.

(7) Find the cubic in x that takes the same values as the function 2^x at the points $x = -1, 0, 1, 2$.

(8) Find the quadratic in x that takes the same values at the points $x = 0, 1, 2$ as the function $1/(x + 2)$. If this quadratic is used as an approximation to $1/(x + 2)$, find the error at $x = \tfrac{1}{2}$.

(9) Form difference tables for the function $\sqrt{x}$ for the range $1 \leqslant x \leqslant 5$, with $h = 1$ and $h = \tfrac{1}{2}$. Compare the columns of second differences in the two tables.

(10) Find the value of the constant k if the quadratic $kx(4 - x)$ takes the same values as the function $\sin(\tfrac{1}{2}\pi x)$ at the points $x = 0, 1, 4$. Show by drawing their graphs that the quadratic is not a good approximation for $\sin(\tfrac{1}{2}\pi x)$ in this range.

Divided differences

If the values of x are not equally spaced, the method above cannot be used, and instead we can employ 'divided differences'. As the name indicates, we define the first divided difference of y_0 and y_1 by

$$[x_0, x_1] = (y_0 - y_1)/(x_0 - x_1) = (y_1 - y_0)/(x_1 - x_0)$$

An alternative notation denotes $[x_0, x_1]$ by $f(x_0, x_1)$.

Our definition gives us the identity

$$y_0 = y_1 + (x_0 - x_1)[x_0, x_1]$$

From three values of y, say y_0, y_1, y_2, we can obtain the second divided difference defined by

$$[x_0, x_1, x_2] = \{[x_0, x_1] - [x_1, x_2]\}/(x_0 - x_2)$$

This gives us the identity

$$[x_0, x_1] = [x_1, x_2] + (x_0 - x_2)[x_0, x_1, x_2]$$

so that we now have

$$y_0 = y_1 + (x_0 - x_1)[x_1, x_2] + (x_0 - x_1)(x_0 - x_2)[x_0, x_1, x_2]$$

The nth divided difference is defined by

$$[x_0, x_1, \ldots, x_n] = \{[x_0, x_1, \ldots, x_{n-1}] - [x_1, x_2, \ldots, x_n]\}/(x_0 - x_n)$$

We can prove by induction the identity

$$y_0 = y_1 + (x_0 - x_1)[x_1, x_2] + (x_0 - x_1)(x_0 - x_2)[x_1, x_2, x_3] + \cdots$$
$$+ (x_0 - x_1)(x_0 - x_2)\ldots(x_0 - x_{n-1})[x_1, x_2, \ldots, x_n]$$
$$+ (x_0 - x_1)(x_0 - x_2)\ldots(x_0 - x_n)[x_0, x_1, x_2, \ldots, x_n]$$

If now we are given the values of y corresponding to the n values $x_1, x_2, \ldots, x_n$ of x we can regard x_0 as a variable and replace x_0, y_0 by x, y.

If $y = x^m$, where m is a positive integer, we have

$$[x, x_1] = (x^m - x_1^m)/(x - x_1) = x^{m-1} + x^{m-2}x_1 + x^{m-3}x_1^2 + \cdots + x_1^{m-1}$$

Thus the first divided difference of y is a polynomial of degree $(m - 1)$. It follows that the divided differences of order m of $y = x^m$ will be 1, and that those of higher order will be zero. In particular, if in the last identity above, y is a polynomial $f(x)$ of degree $(n - 1)$, the final term in the identity is zero and we can write

$$f(x) = y_1 + (x - x_1)[x_1, x_2] + (x - x_1)(x - x_2)[x_1, x_2, x_3] + \cdots$$
$$+ (x - x_1)(x - x_2)\ldots(x - x_{n-1})[x_1, x_2, \ldots, x_n]$$

This gives us the unique polynomial of degree $(n - 1)$ which takes the values $y_1, y_2, \ldots, y_n$ for $x = x_1, x_2, \ldots, x_n$. Thus

$$y - f(x) = (x - x_1)(x - x_2)\ldots(x - x_n)[x, x_1, \ldots, x_n] \qquad (2.1)$$

where y is some function of which we know n values and $f(x)$ is the polynomial above of degree $(n - 1)$ in x. The expression on the right-hand side represents the error incurred if we assume that y and $f(x)$ are equal.

To illustrate this, consider the function $y = 1/x$. Put $x_1 = 2$, $x_2 = 3$, $x_3 = 4$ so that $y_1 = \frac{1}{2}$, $y_2 = \frac{1}{3}$, $y = \frac{1}{4}$. Then we obtain

$$f(x) = \frac{1}{2} + (x - 2)(-\tfrac{1}{6}) + (x - 2)(x - 3)(\tfrac{1}{24})$$

and

$$y - f(x) = (x - 2)(x - 3)(x - 4)[x, 2, 3, 4]$$

It is easy to check that $f(x)$ takes the given values at $x = 2, 3, 4$. Further, if we work out the divided difference $[x, 2, 3, 4]$, we find that it equals $-1/24x$ and that $f(x) - (x - 2)(x - 3)(x - 4)/24x$ simplifies to $1/x$. This emphasizes the fact that the equation (2.1) above expresses an identity.

In order to obtain a more useful form for equation (2.1) we make an assumption about the function y. Usually the only information we have consists of the set of given values. If we now assume that y is continuous and can be differentiated $(n - 1)$ times, we argue that since $f(x)$ (being a polynomial) can also be differentiated $(n - 1)$ times, so can the expression $\{y - f(x)\}$. But this expression is continuous and equals zero at n points in a range of values of x containing the points $x_1, x_2, \ldots, x_n$. It follows that the $(n - 1)$th derivative of $\{y - f(x)\}$ equals zero once in this range, say at some point $x = \xi$. Then

$$y^{(n-1)}(\xi) = (n - 1)![x_1, x_2, \ldots, x_n]$$

since the $(n - 1)$th derivative of $(x - x_1)(x - x_2)\ldots(x - x_{n-1})$ is $(n - 1)!$. This shows that the divided difference $[x_1, x_2, \ldots, x_n]$ is equal to $y^{(n-1)}(\xi)/(n - 1)!$ where ξ is some value of x between the largest and the smallest of $x_1, x_2, \ldots, x_n$. Hence $[x, x_1, x_2, \ldots, x_n]$ is equal to $y^{(n)}(\xi)/n!$, where a suitable value for ξ is chosen between the largest and the smallest of $x, x_1, x_2, \ldots, x_n$. Equation (2.1) can now be written in the form

$$y = f(x) + (x - x_1)(x - x_2)\ldots(x - x_n)y^{(n)}(\xi)/n!$$

If we can glean any information about the magnitude of the nth derivative of y, this equation helps us to estimate the error we make in assuming that y and $f(x)$ are equal.

In the example above we had

$$y = 1/x, \qquad f(x) = \tfrac{1}{2} + (x - 2)(-\tfrac{1}{6}) + (x - 2)(x - 3)(\tfrac{1}{24})$$

If we put $x = 2 \cdot 6$ we find

$$y = 0 \cdot 3846, \qquad f(x) = 0 \cdot 39, \qquad y - f(x) = -0 \cdot 0054$$

Also

$$(x - 2)(x - 3)(x - 4)y^{(3)}(\xi)/6 = -(0 \cdot 336)/\xi^4$$

This gives $\xi^4 = (0 \cdot 336)/(0 \cdot 0054) = 62$ approximately, so that the value of ξ lies in the range containing the points $x = 2, 2 \cdot 6, 3, 4$.

We define $[x_1, x_1]$ as the limit of $[x_1, x_2]$ as x_2 tends to x_1, so that

$$[x_1, x_1] = \lim_{x_2 \to x_1} \frac{(y_1 - y_2)}{(x_1 - x_2)} = y_1'$$

Similarly

$$[x_1, x_2, x_2] = \lim_{x_3 \to x_2} \frac{[x_1, x_2] - [x_1, x_3]}{(x_2 - x_3)} = \frac{d}{dx}[x_1, x] \quad \text{at } x = x_2$$

It follows that the derivative of $[x_1, x]$ is $[x_1, x, x]$, and the derivative of $[a, b, c, x]$ is $[a, b, c, x, x]$.

EXAMPLES

(1) Find the polynomial $y(x)$ of degree 4 which takes the following values:

$$x \quad 1 \quad 2 \quad 0 \quad -1 \quad 4$$

$$y \quad -15 \quad 0 \quad -16 \quad -15 \quad 240$$

Find also a polynomial of degree 5 which takes these values.
We form the table of divided differences:

$$x_1 = \quad 1, y_1 = -15$$
$$15$$
$$x_2 = \quad 2, y_2 = \quad 0 \qquad 7$$
$$8 \qquad 2$$
$$x_3 = \quad 0, y_3 = -16 \qquad 3 \qquad 1$$
$$-1 \qquad 5$$
$$x_4 = -1, y_4 = -15 \qquad 13$$
$$51$$
$$x_5 = \quad 4, y_5 = \quad 240$$

For example,

$$[x_4, x_5] = (-15 - 240)/(-1 - 4) = 51$$
$$[x_3, x_4, x_5] = (-1 - 51)/(0 - 4) = 13$$
$$[x_2, x_3, x_4, x_5] = (3 - 13)/(2 - 4) = 5$$
$$[x_1, x_2, x_3, x_4, x_5] = (2 - 5)/(1 - 4) = 1$$

The required polynomial is

$$y_1 + (x - x_1)[x_1, x_2] + (x - x_1)(x - x_2)[x_1, x_2, x_3]$$
$$+ (x - x_1)(x - x_2)(x - x_3)[x_1, x_2, x_3, x_4]$$
$$+ (x - x_1)(x - x_2)(x - x_3)(x - x_4)[x_1, x_2, x_3, x_4, x_5]$$
$$= -15 + 15(x - 1) + 7(x - 1)(x - 2) + 2x(x - 1)(x - 2)$$
$$+ x(x^2 - 1)(x - 2)$$

This simplifies to $(x^4 - 16)$, which takes the values given. Consider now the expression

$$(x^4 - 16) + k(x - 1)(x - 2)(x)(x + 1)(x - 4)$$

This is a polynomial of degree 5 which takes the required values for any value of the constant k.

(2) If $y(x)$ is a cubic polynomial such that $y(1) = -2, y(3) = 16, y(-2) = 1$, $y(-3) = -14$, find the value of $y(2)$ without finding the polynomial.

Form the table of divided differences:

$$
\begin{array}{ll}
x & y \\
1 & -2 \\
 & \qquad (-2 - 16)/(1 - 3) = 9 \\
3 & 16 \qquad\qquad\qquad\qquad (9 - 3)/(1 + 2) = 2 \\
 & \qquad (16 - 1)/(3 + 2) = 3 \qquad\qquad\qquad (2 + 2)/(1 + 3) = 1 \\
-2 & 1 \qquad\qquad\qquad\qquad (3 - 15)/(3 + 3) = -2 \\
 & \qquad (1 + 14)/(-2 + 3) = 15 \qquad\qquad [3, -2, -3, 2] \\
-3 & -14 \qquad\qquad\qquad\qquad [-2, -3, 2] \\
 & \qquad [-3, 2] \\
2 & y(2)
\end{array}
$$

Since y is a cubic polynomial, the third-order divided differences are constant, so that $[3, -2, -3, 2]$ equals 1.

Now $\quad \{[3, -2, -3] - [-2, -3, 2]\}/(3 - 2) = [3, -2, -3, 2]$

i.e., $\qquad -2 - [-2, -3, 2] = 1, \qquad [-2, -3, 2] = -3$

Next we have

$$\{[-2, -3] - [-3, 2]\}/(-2 - 2) = [-2, -3, 2] = -3$$

i.e., $\qquad 15 - [-3, 2] = 12, \qquad [-3, 2] = 3$

Finally $\qquad \{-14 - y(2)\}/(-3 - 2) = [-3, 2] = 3$

and hence $\qquad\qquad\qquad\qquad y(2) = 1$

To check this result we can take the same values of x in a different order, form the difference table and show that the third-order differences equal 1.

(3) Find the quadratics $y(x)$ that take the following sets of values:

$$
\begin{array}{llll}
(a) & y(1) = 2, & y(3) = 12, & y(4) = 20 \\
(b) & y(-1) = 0, & y(1) = 0, & y(2) = 6 \\
(c) & y(-2) = 0, & y(1) = -3, & y(2) = 0 \\
(d) & y(1) = 1, & y(4) = 10, & y(2) = 2
\end{array}
$$

(4) Form the divided difference table for the function $y(x)$ that takes the following values. Assuming that the function is a polynomial of degree 4 in x, use the table to calculate $y(2)$ and $y(3)$.

$$y(-3) = 36, \qquad y(-2) = 1, \qquad y(0) = 9, \qquad y(1) = 4, \qquad y(4) = 169$$

(5) A function $y(x)$ takes the following values:

$$y(x_0) = 7, \qquad y(x_1) = 6, \qquad y(x_2) = 15, \qquad y(x_3) = -20$$

Calculate the divided differences $[x_0, x_1, x_2]$, $[x_2, x_0, x_1]$, $[x_0, x_1, x_2, x_3]$, $[x_2, x_1, x_3, x_0]$. If $x_0 = 0$, $x_1 = 1$, $x_2 = -2$, $x_3 = 3$, find $y(2)$ on the assumption that y is a cubic in x.

(6) If $y = ax + b$, show that the divided difference $[x_0, x_1]$ does not depend on x_0 or x_1.

(7) If $y = x^4$ find

    (a) the limit of $[x_0, x_1]$ as x_1 tends to x_0
    (b) the limit of $[x_0, x_1, x_2]$ as x_2 tends to x_1
    (c) the limit of $[x_0, x_1, x_1]$ as x_1 tends to x_0

(8) Form the divided difference table for $y = x^4 - x^3$ using the points $x = 0, 2, -1, 3, -2$ in that order and verify that the fourth divided difference is unity. Repeat with the points in a different order.

(9) Show that for the function y the divided difference $[a, b, c]$ equals

$$y(a)/(a - b)(a - c) + y(b)/(b - a)(b - c) + y(c)/(c - a)(c - b)$$

(10) Show that the derivative of $[a, x, x]$ is $2[a, x, x, x]$.

First-order difference equations

Our aim here is to express y_n in terms of n, starting from an equation for Δy_n, with $h = 1$.

    (a) If $\Delta y_n = a$, then $y_{n+1} - y_n = a$. This recurrence relation gives $y_{n+1} = y_1 + na$, or $y_n = y_0 + na$.

    (b) If $\Delta y_n = ay_n$, we have $y_{n+1} = (a + 1)y_n = (a + 1)^n y_1$. If $\Delta y_n = ay_n + b$, we can put $z_n = y_n + b/a$.

    (c) If $\Delta y_n = na$, then $y_{n+1} = y_n + na$, which leads to

$$y_{n+1} = y_1 + \tfrac{1}{2}n(n + 1)a, \qquad y_n = y_0 + \tfrac{1}{2}n(n - 1)a$$

    (d) If $\Delta y_n = a^n$ we have $y_{n+1} = y_n + a^n = y_1 + a^n + a^{n-1} + \cdots + a$, so that

$$y_n = y_0 + (1 - a^n)/(1 - a)$$

EXAMPLES

(1) Solve $\Delta y_n = 2$, given that $y_1 = 1$ $(y_n = 2n - 1)$
(2) Solve $\Delta y_n = 2y_n$, given that $y_1 = 2$ $(y_n = 2 \cdot 3^{n-1})$
(3) Solve $\Delta y_n = 2y_n + 4$, given that $y_0 = 0$ $(y_n = 2 \cdot 3^n - 2)$
(4) Solve $\Delta y_n = 2n + 3$, given that $y_0 = 1$ $(y_n = n^2 + 2n + 1)$
(5) Solve $\Delta y_n = 3^n$, given that $y_0 = 4$ $(2y_n = 3^n + 7)$

Second-order difference equations

Consider first the equation $\Delta^2 y_n = 0$. This equation holds when $y_n = An + B$. We can compare this with $y = Ax + B$, the solution of the differential equation $y'' = 0$. Since $\Delta = E - 1$, the general second-order difference equation

$$\Delta^2 y_n + a\Delta y_n + by_n = f(n)$$

can be written as the recurrence relation

$$E^2 y_n - (2 - a)E y_n + (1 - a + b)y_n = f(n)$$

i.e.,
$$y_{n+2} - (2 - a)y_{n+1} + (1 - a + b)y_n = f(n)$$

Consider now the equation

$$y_{n+2} + p y_{n+1} + q y_n = 0$$

The type of solution depends on the nature of the roots of the quadratic equation $x^2 + px + q = 0$.

(a) Real unequal roots

The equation
$$y_{n+2} - 5y_{n+1} + 6y_n = 0$$

can be written
$$(E^2 - 5E + 6)y_n = 0$$

or
$$(E - 2)(E - 3)y_n = 0$$

or
$$(E - 3)(E - 2)y_n = 0$$

This equation is satisfied if either $(E - 3)y_n = 0$, or $(E - 2)y_n = 0$, i.e., if $y_{n+1} = 3y_n$ or $2y_n$, so that $y_n = 3^n y_0$ or $2^n y_0$. Since the equation is linear, the general solution is $y_n = A \cdot 3^n + B \cdot 2^n$. If we know the values of y_0 and y_1 we can find the constants A and B.

(b) Equal roots

The equation
$$y_{n+2} - 6y_{n+1} + 9y_n = 0$$

gives
$$(E - 3)(E - 3)y_n = 0$$

Clearly $y_n = 3^n y_0$ is one solution.

To find the general solution we substitute $y_n = 3^n z_n$.

Then
$$(E - 3)y_n = 3^{n+1}(E - 1)z_n = 3^{n+1}\Delta z_n$$
$$(E - 3)(E - 3)y_n = 3^{n+2}(E - 1)\Delta z_n$$
$$= 3^{n+2}\Delta^2 z_n$$

But $\Delta^2 z_n = 0$ gives $z_n = An + B$, and so $y_n = 3^n(An + B)$.

(c) Complex roots

Take first the simplest case $y_{n+2} + y_n = 0$. We need a function of n that changes sign when n is increased by 2. This requirement is met by $\cos\left(\tfrac{1}{2}n\pi\right)$ and by $\sin\left(\tfrac{1}{2}n\pi\right)$, so that the general solution is

$$y_n = A \cos\left(\tfrac{1}{2}n\pi\right) + B \sin\left(\tfrac{1}{2}n\pi\right)$$

in which $A = y_0$ and $B = y_1$. Similarly, the equation

$$y_{n+2} + k^2 y_n = 0$$

is satisfied by

$$y_n = k^n\{A \cos\left(\tfrac{1}{2}n\pi\right) + B \sin\left(\tfrac{1}{2}n\pi\right)\}$$

In the general case, put $k = \sqrt{q}$, $2k \cos \theta = -p$, $2k \sin \theta = \sqrt{(4q - p^2)}$. The difference equation then becomes

$$y_{n+2} - (2k \cos \theta)y_{n+1} + k^2 y_n = 0$$

This is satisfied by $y_n = k^n \cos n\theta$ and by $y_n = k^n \sin n\theta$,

since $\qquad \cos (n + 2)\theta - 2 \cos \theta \cos (n + 1)\theta + \cos n\theta = 0$

and $\qquad \sin (n + 2)\theta - 2 \cos \theta \sin (n + 1)\theta + \sin n\theta = 0$

The general solution will be

$$y_n = k^n(A \cos n\theta + B \sin n\theta)$$

For example, the equation $y_{n+2} - 5y_{n+1} + 25y_n = 0$ is satisfied by

$$y_n = 5^n\{A \cos (n\pi/3) + B \sin (n\pi/3)\}$$

and the equation $y_{n+2} - 6y_{n+1} + 25y_n = 0$ is satisfied by

$$y_n = 5^n(A \cos n\theta + B \sin n\theta)$$

where $\cos \theta = 0.6$, $\sin \theta = 0.8$.

We now summarize these results for the difference equation

$$y_{n+2} + py_{n+1} + qy_n = 0$$

If $\qquad p^2 > 4q, \qquad y_n = A\alpha^n + B\beta^n$

where α and β are the roots of the quadratic $x^2 + px + q = 0$.

If $\qquad p^2 = 4q, \qquad y_n = k^n(An + B), \quad$ where $k = -\tfrac{1}{2}p$

If $\qquad p^2 < 4q, \qquad y_n = k^n(A \cos n\theta + B \sin n\theta)$

where $k = \sqrt{q}$ and $\cos \theta = -p/2k$, $\sin \theta = \sqrt{(1 - p^2/4q)}$.

There is a close resemblance between these results and those for the differential equation $y'' + py' + qy = 0$. Thus:

If $\qquad p^2 > 4q, \qquad y = A\,e^{\alpha x} + B\,e^{\beta x}$

If $\qquad p^2 = 4q, \qquad y = e^{-px/2}\,(Ax + B)$

If $\qquad p^2 < 4q, \qquad y = e^{-px/2}\,(A \cos mx + B \sin mx)$

where $\qquad m = \tfrac{1}{2}\sqrt{(4q - p^2)}$

We shall see later how difference equations are used in the numerical solution of differential equations.

EXAMPLES

(1) Solve $y_{n+2} + 5y_{n+1} + 6y_n = 0$ given $y_0 = y_1 = 1$.

The roots of the equation $x^2 + 5x + 6 = 0$ are -2 and -3. Hence

$$y_n = A(-2)^n + B(-3)^n$$

Put $n = 0$, $\qquad\qquad y_0 = A + B = 1$

Put $n = 1$, $y_1 = -2A - 3B = 1$

Hence $A = 4$, $B = -3$, $y_n = 4(-2)^n - 3(-3)^n$

(2) Solve $y_{n+2} + 8y_{n+1} + 16y_n = 0$, given $y_0 = 1, y_1 = 0$.

Since the equation $x^2 + 8x + 16 = 0$ has equal roots and gives $x = -4$,

$$y_n = (-4)^n(An + B)$$

From $n = 0$, $B = 1$, and then from $n = 1$, $A = -1$, and so

$$y_n = (-4)^n(-n + 1)$$

(3) Solve $y_{n+2} + 16y_n = 0$, given $y_0 = 2, y_1 = 4$.

We consider the functions $4^n \cos n\pi/2$ and $4^n \sin n\pi/2$. If n is increased to $n + 2$, each is multiplied by -16. Hence

$$y_n = (4^n)(A \cos n\pi/2 + B \sin n\pi/2)$$

From $y_0 = 2$, we have $A = 2$, and from $y_1 = 4$, we have $B = 1$.

Hence $y_n = (4^n)(2 \cos n\pi/2 + \sin n\pi/2)$

(4) Solve the difference equation $\Delta^2 y_n + 8y_n = 0$, given $y_0 = y_1 = 2$.

We can put this in the form

$$(E - 1)^2 y_n + 8y_n = y_{n+2} - 2y_{n+1} + 9y_n = 0$$

Here $k = 3$, and by substituting $y_n = 3^n \cos n\theta$, we obtain

$$9 \cos (n + 2)\theta - 6 \cos (n + 1)\theta + 9 \cos n\theta = 0$$

i.e., $6 \cos (n + 1)\theta\{3 \cos \theta - 1\} = 0$

$$\cos \theta = \tfrac{1}{3}, \qquad \theta = 70° 32'$$

$$y_n = 3^n(A \cos n\theta + B \sin n\theta)$$

From $y_0 = 2$, we find $A = 2$, and from $y_1 = 2$, we find $B = 0$. The solution is therefore $y_n = 2.3^n \cos n\theta$.

(5) Solve $y_{n+2} - 9y_n = 0$, given $y_0 = 2, y_1 = 4$.

(6) Solve $y_{n+2} - 4y_{n+1} + 4y_n = 0$, given $y_0 = y_1 = 3$.

(7) Solve $y_{n+2} + 9y_n = 0$, given $y_0 = 0, y_1 = 3$.

(8) Solve $y_{n+2} + y_{n+1} + y_n = 0$, given $y_0 = 1, y_1 = 2$.

(9) Solve the equation $\Delta^2 y_n - 6\Delta y_n + 18y_n = 0$, with $y_0 = 0, y_1 = 3$.

Equations of the form $y_{n+2} + py_{n+1} + qy_n = f(n)$

As with the second-order differential equation $y'' + py' + qy = f(x)$, the solution is the sum of the complementary function and a particular solution. When these are substituted in $(y_{n+2} + py_{n+1} + qy_n)$, the complementary function produces zero, while any particular solution will produce $f(n)$.

Take first the example

$$y_{n+2} - 3y_{n+1} + 2y_n = 2n$$

The complementary function is $A(1^n) + B(2^n)$.

Now we can rewrite our equation in the form $\Delta^2 y_n - \Delta y_n = 2n$. Put $z_n = \Delta y_n$, so that $(\Delta - 1)z_n = 2n$. This gives

$$\Delta y_n = -(1 - \Delta)^{-1}2n = -(1 + \Delta + \Delta^2 + \cdots)2n = -2n - 2$$

Hence a particular solution is $y_n = -n^2 - n$, so that the final result is $y_n = A + B.2^n - n^2 - n$.

Now consider the equation

$$3y_{n+2} + 4y_{n+1} + y_n = 32n^2 + 48n$$

The complementary function is $A(-1)^n + B(-\tfrac{1}{3})^n$. We have

$$3y_{n+2} + 4y_{n+1} + y_n = (3\Delta^2 + 10\Delta + 8)y_n$$

Thus
$$(8 + 10\Delta + 3\Delta^2)y_n = 32n^2 + 48n$$

We write this in the form

$$\begin{aligned}
y_n &= (8 + 10\Delta + 3\Delta^2)^{-1}(32n^2 + 48n) \\
&= 3(1 + \tfrac{3}{4}\Delta)^{-1}(4n^2 + 6n) - 2(1 + \tfrac{1}{2}\Delta)^{-1}(4n^2 + 6n) \\
&= (1 - \tfrac{3}{4}\Delta + \tfrac{9}{16}\Delta^2 - \cdots)(12n^2 + 18n) \\
&\quad -(1 - \tfrac{1}{2}\Delta + \tfrac{1}{4}\Delta^2 - \cdots)(8n^2 + 12n) \\
&= (12n^2 - 9) - (8n^2 + 4n - 6) = 4n^2 - 4n - 3
\end{aligned}$$

We now have the solution

$$y_n = A(-1)^n + B(-\tfrac{1}{3})^n + 4n^2 - 4n - 3$$

We could instead assume that the particular solution would be of the form $an^2 + bn + c$. Then

$$(8 + 10\Delta + 3\Delta^2)(an^2 + bn + c) = 8an^2 + (20a + 8b)n + (16a + 10b + 8c)$$

Comparison with $32n^2 + 48n$ gives $a = 4$, $b = -4$, $c = -3$.

We deal similarly with the equation

$$y_{n+2} + y_{n+1} + 4y_n = 3^n$$

The complementary function is $2^n(A \cos n\theta + B \sin n\theta)$ where $\cos \theta = -\tfrac{1}{4}$, $\sin \theta = \tfrac{1}{4}\sqrt{15}$.

We assume that a particular solution will be given by a constant multiple of 3^n. If we put $y_n = 3^n$, then

$$y_{n+2} + y_{n+1} + 4y_n = (9 + 3 + 4)3^n = 16.3^n$$

Hence a particular solution is $3^n/16$, and the general solution is

$$y_n = (2^n)(A \cos n\theta + B \sin n\theta) + 3^n/16$$

EXAMPLES

(1) Solve $y_{n+2} + 3y_{n+1} + 9y_n = 13n^2 + 10n + 7$.

(2) Solve $y_{n+2} + 6y_{n+1} + 5y_n = 12n + 20$.

(3) Solve $y_{n+2} - y_{n+1} + 6y_n = 2^n$.

(4) The solution of differential equations by the Adams–Bashforth method involves the difference equation

$$(1 - 5a)y_{n+2} - (1 + 8a)y_{n+1} + ay_n = 0$$

If a is small show that the general solution is approximately

$$y_n = A(1 + 12a)^n + Ba^n$$

(5) In Milne's method for the solution of differential equations, the stability of the solution depends on the difference equation

$$(1 - a)y_{n+2} - 4ay_{n+1} - (1 + a)y_n = 0$$

If a is small, show that the general solution is approximately

$$y_n = A(1 + 3a)^n + B(-1 + a)^n$$

(6) Solve $y_{n+3} - 3y_{n+2} - 4y_{n+1} + 12y_n = 0$ given that $y_0 = 0$, $y_1 = 6$, $y_2 = 10$. Check by finding y_3 and y_4 from your solution and from the equation.

(7) Show that the equation $y_{n+2} - 6y_{n+1} + 25y_n = 0$ is satisfied by

$$y_n = \alpha(3 - 4i)^n + \beta(3 + 4i)^n$$

where α and β are conjugate complex constants. Deduce the solution given above.

3

Interpolation

If we know the values of the function $y(x)$ for a set of values of x, we can use these values to estimate the value of y at a fresh value of x. Thus if $y(0) = 0$, $y(1) = 2$, $y(2) = 4$, and $y(4) = 8$, it seems reasonable to assume that $y(3) = 6$. This will be so if $y(x) = 2x$, but it will not be true if for example

$$y = 2x + x(x - 1)(x - 2)(x - 4)$$

or

$$y = 2 \sin (\pi x/2) + x \cos \pi x + x$$

or

$$y = 2^x - (x - 1)(x - 2)(12 + 5x)/24$$

Each of these three functions takes the required values at $x = 0, 1, 2, 4$ but none of them gives $y(3) = 6$.

It is clear that we cannot place any reliance upon a value of y estimated on such scanty information as above. To be certain that our estimate is correct we should need sufficient information, e.g., that $y(5) = -2$ and that y is a polynomial in x of degree 4. We could then show that

$$y = 2x - x(x - 1)(x - 2)(x - 4)/5$$

and so $y(3) = 7 \cdot 2$.

Linear interpolation is based on the equation

$$y_p = y_0 + p\Delta y_0 = (1 - p)y_0 + py_1$$

which is accurate if the second differences can be neglected. As a simple example of linear interpolation, we consider finding the value of sec 68° 9′ from four-figure tables. These give

$$\sec 68° \ 6′ = 2 \cdot 6811, \qquad \sec 68° \ 12′ = 2 \cdot 6927$$

In the mean difference column for 3′, we find the entry $0 \cdot 0060$, and this gives an answer 2·6871. This is incorrect. The increment for 3 minutes varies from 0·0058 to 0·0063 between 68° and 69°, and the increment 0·0060 is an average (or mean) value. If $y_0 = 2·6811$ and $y_1 = 2·6927$ we find $\Delta y_0 = 0·0116$. Then with $p = \frac{1}{2}$,

$$y_{\frac{1}{2}} = E^{\frac{1}{2}}y_0 = (1 + \Delta)^{\frac{1}{2}}y_0 = (1 + \tfrac{1}{2}\Delta - \tfrac{1}{8}\Delta^2 + \cdots)y_0$$

$$y_{\frac{1}{2}} = 2·6811 + \tfrac{1}{2}(0·0116) = 2·6869$$

(as $\frac{1}{8}\Delta^2 y_0$ is negligible). This is correct to four decimal places.

Quadratic interpolation is based on the equation

$$y_p = y_0 + p\Delta y_0 + \tfrac{1}{2}p(p - 1)\Delta^2 y_0$$

For an example of quadratic interpolation, we evaluate cosh 1·305 from the difference table below:

x	$\cosh x$		
1·30	1·9709		
		344	
1·32	2·0053		7
		351	
1·34	2·0404		

Again we have

$$y_p = (1 + \Delta)^p y_0 = y_0 + p\Delta y_0 + \tfrac{1}{2}p(p - 1)\Delta^2 y_0 + \cdots$$

Since $h = 0·02$, we put $p = \frac{1}{4}$ and obtain

$$\cosh 1·305 = 1·9709 + \tfrac{1}{4}(0·0344) - \tfrac{3}{32}(0·0007) = 1·9794$$

In this case we can express y_p in terms of y_0, y_1, and y_2.

$$y_p = y_0 + p(y_1 - y_0) + \tfrac{1}{2}p(p - 1)(y_2 - 2y_1 + y_0)$$
$$= \tfrac{1}{2}(p - 1)(p - 2)y_0 - p(p - 2)y_1 + \tfrac{1}{2}p(p - 1)y_2$$

It can be seen that this formula gives the values y_0, y_1, y_2 for $p = 0, 1, 2$ respectively. In both the previous examples there is some uncertainty about the final digit in the answer because of round-off error in the data.

EXAMPLES

(1) Find tan 78° 57′ and tan 78° 58′ by interpolation from tables.

(2) From the following table of values calculate e^x for $x = 2·01, 2·03, 2·05$.

x	2·00	2·02	2·04	2·06
e^x	7·3891	7·5383	7·6906	7·8460

Compare your results with a printed table.

(3) Calculate the value of $x^3 - 6x^2 + 12x$ at $x = 1, 2, 3$. Using these

values, find by interpolation the estimated value at $x = 1\cdot5$ and compare it with the value of $x^3 - 6x^2 + 12x$ at this point.

(4) Show that if third-order differences are negligible

$$y_p = y_0 + p\nabla y_0 + \tfrac{1}{2}p(p + 1)\nabla^2 y_0$$

Calculate $\tan 45^\circ\,24'$ and $\tan 44^\circ\,36'$, given that $\tan 43^\circ = 0\cdot9325$, $\tan 44^\circ = 0\cdot9657$, and $\tan 45^\circ = 1$.

The Newton forward difference formula

Our starting point is the divided difference formula

$$y(x) = f(x) + (x - x_1)(x - x_2)\ldots(x - x_n)y^{(n)}(\xi)/n!$$

where

$$f(x) = y_1 + (x - x_1)[x_1, x_2] + (x - x_1)(x - x_2)[x_1, x_2, x_3]$$
$$+ \cdots + (x - x_1)(x - x_2)\ldots(x - x_{n-1})[x_1, x_2, \ldots, x_n]$$

Here we know the value of y at the n points $x = x_1, x_2, \ldots, x_n$, and if the value of

$$(x - x_1)(x - x_2)\ldots(x - x_n)y^{(n)}(\xi)/n!$$

is negligible, we can by evaluating $f(x)$ obtain the value of $y(x)$ at a new point x. Notice however that the choice of x has some bearing on the value of ξ, which was shown to be a point in an interval containing the points $x, x_1, x_2, \ldots, x_n$.

As it stands, this formula holds good whether the values of x are equally spaced or not. We now restrict ourselves to the case when $x_r = x_0 + rh$, giving equal spacing, and this enables us to express divided differences in terms of forward differences.

$$[x_1, x_2] = (y_1 - y_2)/(x_1 - x_2) = \Delta y_1/h$$
$$[x_1, x_2, x_3] = \{[x_3, x_2] - [x_2, x_1]\}/2h = (\Delta y_2 - \Delta y_1)/2h^2 = \Delta^2 y_1/2h^2$$

By induction we can show that $[x_1, x_2, \ldots, x_{r+1}] = \Delta^r y_1/(r!)h^r$.

Next we put $x = x_1 + ph$, so that

$$(x - x_1)(x - x_2) = p(p - 1)h^2$$
$$(x - x_1)(x - x_2)(x - x_3) = p(p - 1)(p - 2)h^3 \quad \text{and so on}$$

The formula now becomes

$$y(x) = y_1 + p\Delta y_1 + p(p - 1)\Delta^2 y_1/2! + p(p - 1)(p - 2)\Delta^3 y_1/3! + \cdots$$
$$+ p(p - 1)\ldots(p - n + 2)\Delta^{n-1} y_1/(n - 1)!$$
$$+ p(p - 1)\ldots(p - n + 1)h^n y^{(n)}(\xi)/n!$$

We now use y_p to denote $y(x)$, and obtain the Newton forward difference formula if we name our points $x_0, x_1, \ldots, x_{n-1}$ in place of $x_1, x_2, \ldots, x_n$.

This gives

$$y_p = y_0 + p\Delta y_0 + p(p-1)\Delta^2 y_0/2! + p(p-1)(p-2)\Delta^3 y_0/3! + \cdots$$
$$+ p(p-1)\ldots(p-n+2)\Delta^{n-1}y_0/(n-1)!$$
$$+ p(p-1)\ldots(p-n+1)h^n y^{(n)}(\xi)/n!$$

The last term is known as the remainder term and plays the same role as the corresponding term in a Taylor series. The formal expansion of $(1 + \Delta)^p y_0$ will give the Newton formula, but it fails to give the remainder term. It is useful as it stands if we are told that y is a polynomial of degree m, say, for then the remainder term is zero when $n = m + 1$, and we have exactly

$$y_p = y_0 + p\Delta y_0 + \tfrac{1}{2}p(p-1)\Delta^2 y_0 + \cdots + p(p-1)\ldots(p-m+1)\Delta^m y_0/m!$$

EXAMPLE

$y(x)$ is a polynomial of degree 4 such that $y(0) = 6, y(1) = y(2) = 5, y(3) = 15, y(4) = 50$. Find $y(\tfrac{1}{2})$.

We form the forward difference table

$$
\begin{array}{ccccccccc}
6 \\
 & -1 = \Delta y_0 \\
5 & & 1 = \Delta^2 y_0 \\
 & 0 & & 9 = \Delta^3 y_0 \\
5 & & 10 & & 6 = \Delta^4 y_0 \\
 & 10 & & 15 \\
15 & & 25 \\
 & 35 \\
50
\end{array}
$$

Then
$$y_{\frac{1}{2}} = 6 + \tfrac{1}{2}(-1) - \tfrac{1}{8}(1) + \tfrac{1}{16}(9) - \tfrac{5}{128}(6) = 5\tfrac{45}{64}$$

To obtain a check on this, we employ the same formula starting with y_1, taking $p = -\tfrac{1}{2}$. As we know that $\Delta^4 y_1 = 6$, this gives

$$y_{\frac{1}{2}} = 5 - \tfrac{1}{2}(0) + \tfrac{3}{8}(10) - \tfrac{5}{16}(15) + \tfrac{35}{128}(6) = 5\tfrac{45}{64}$$

The Newton forward difference formula is used normally with values of p between 0 and 1. The series obtained usually converges slowly, and if p is taken greater than 1 the rate of convergence is slower still.

The influence of the size of the step-length h on the rate of convergence can be illustrated by considering the function $y = e^x$ with $x_r = rh$. We have $\Delta^n y_0 = (e^h - 1)^n$, so that in the Newton formula the ratio of $(r + 1)$th term to the rth term is $(p - r + 1)(e^h - 1)/r$. If e^h is greater than 2, i.e., if $h > 0.7$, the series will not converge at all. If we take a step-length equal to 0.1, $(e^h - 1)$ will equal 0.105, and the series will converge fairly quickly. In practice, the number of decimal places used and the size of round-off errors will be important factors affecting the choice of h. The Newton formula suffers from the defect that the given values of y which are employed are all (except for y_0) for values of x above x_0. This is a lopsided arrangement. To obtain a better balanced arrangement we return to the divided difference formula.

The Gauss forward formula

We employ the values of y at the $(2n + 1)$ points $x_0, x_1, x_{-1}, x_2, x_{-2}, \ldots,$ x_n, x_{-n}, and these points enter the formula in this order. We first express the divided differences as central differences.

$$[x_0, x_1] = \delta y_{\frac{1}{2}}/h$$
$$[x_0, x_1, x_{-1}] = [x_1, x_0, x_{-1}] = (\delta y_{\frac{1}{2}} - \delta y_{-\frac{1}{2}})/2h^2 = \delta^2 y_0/2h^2$$
$$[x_0, x_1, x_{-1}, x_2] = [x_2, x_1, x_0, x_{-1}] = \delta^3 y_{\frac{1}{2}}/(3!)h^3$$

We also put $x - x_0 = ph$, $x - x_r = (p - r)h$, so that we obtain

$$y_p = y_0 + p\delta y_{\frac{1}{2}} + \tfrac{1}{2}p(p - 1)\delta^2 y_0 + (p + 1)p(p - 1)\delta^3 y_{\frac{1}{2}}/3!$$
$$+ (p + 1)p(p - 1)(p - 2)\delta^4 y_0/4! + \cdots$$

We can write the formula in the more concise form

$$y_p = y_0 + p\delta y_{\frac{1}{2}} + \binom{p}{2}\delta^2 y_0 + \binom{p + 1}{3}\delta^3 y_{\frac{1}{2}} + \binom{p + 1}{4}\delta^4 y_0 + \cdots$$

$$+ \binom{p + n - 1}{2n}\delta^{2n} y_0 + \binom{p + n}{2n + 1}h^{2n + 1}y^{(2n + 1)}(\xi)$$

where $x_{-n} < \xi < x_n$.

The Gauss backward formula

In the same way, using the same points in the order $x_0, x_{-1}, x_1, x_{-2}, x_2, \ldots,$ x_{-n}, x_n, we obtain

$$y_p = y_0 + p\delta y_{-\frac{1}{2}} + \binom{p + 1}{2}\delta^2 y_0 + \binom{p + 1}{3}\delta^3 y_{-\frac{1}{2}} + \binom{p + 2}{4}\delta^4 y_0 + \cdots$$

$$+ \binom{p + n}{2n}\delta^{2n} y_0 + \binom{p + n}{2n + 1}h^{2n + 1}y^{(2n + 1)}(\xi)$$

The value of ξ here will be exactly the same as in the Gauss forward formula, as the two formulae are based on the same set of $(2n + 1)$ values. These two formulae will produce the same answer if we truncate them both after an even-order difference.

4

Stirling's formula

This is used especially for the range of values $-\frac{1}{4} < p < \frac{1}{4}$. We obtain it at once by taking the mean of the two Gauss formulae.

$$y_p = y_0 + p\mu\delta y_0 + p^2\delta^2 y_0/2! + p(p^2 - 1)\mu\delta^3 y_0/3!$$
$$+ p^2(p^2 - 1)\delta^4 y_0/4! + p(p^2 - 1)(p^2 - 4)\mu\delta^5 y_0/5! + \cdots$$

where $\mu\delta y_0 = \frac{1}{2}(\delta y_{\frac{1}{2}} + \delta y_{-\frac{1}{2}})$ and $\mu\delta^3 y_0 = \frac{1}{2}(\delta^3 y_{\frac{1}{2}} + \delta^3 y_{-\frac{1}{2}})$.

EXAMPLE

Construct a difference table for the values to six decimal places of $\sin x°$ for $x = 0, 5, 10, 15, 20, 25, 30$. Use Stirling's formula to calculate $\sin 14°$ and $\sin 16°$.

Bessel's formula

To derive a formula symmetrical about the mid-point of the interval $x_0 < x < x_1$, we make the substitutions

$$\delta^2 y_0 = \frac{1}{2}(\delta^2 y_0 + \delta^2 y_1) - \frac{1}{2}\delta^3 y_{\frac{1}{2}}$$
$$\delta^4 y_0 = \frac{1}{2}(\delta^4 y_0 + \delta^4 y_1) - \frac{1}{2}\delta^5 y_{\frac{1}{2}}$$

in the Gauss forward formula. This gives us Bessel's formula

$$y_p = y_0 + p\delta y_{\frac{1}{2}} + B_2(\delta^2 y_0 + \delta^2 y_1) + B_3\delta^3 y_{\frac{1}{2}}$$
$$+ B_4(\delta^4 y_0 + \delta^4 y_1) + B_5\delta^5 y_{\frac{1}{2}} + \cdots$$

where
$$B_2 = p(p - 1)/4$$
$$B_3 = p(p - \tfrac{1}{2})(p - 1)/6$$
$$B_4 = (p + 1)p(p - 1)(p - 2)/48$$
$$B_5 = (p + 1)p(p - \tfrac{1}{2})(p - 1)(p - 2)/120$$

This formula has no superior for interpolation in the range $\frac{1}{4} < p < \frac{3}{4}$. Tables of the values of the coefficients B_2, B_3, B_4, B_5 are obtainable, e.g., in *Interpolation and Allied Tables*, H.M.S.O. In this connection it is worthwhile to quote an extract from a critical table for B_2

p	B_2
0·3904	
	−0·060
0·4105	
	−0·061
0·4367	

i.e.,
$$B_2 = -0·060 \quad \text{if} \quad 0·3904 < p \leqslant 0·4105$$
$$B_2 = -0·061 \quad \text{if} \quad 0·4105 < p \leqslant 0·4367$$

Everett's formula

To obtain this we substitute

$$\delta^3 y_{\frac{1}{2}} = \delta^2 y_1 - \delta^2 y_0, \qquad \delta^5 y_{\frac{1}{2}} = \delta^4 y_1 - \delta^4 y_0$$

in the Gauss forward formula. We obtain a formula in which $\delta y_{\frac{1}{2}}$ is the only odd order difference.

$$\begin{aligned}
y_p &= y_0 + p\delta y_{\frac{1}{2}} + E_2\delta^2 y_0 + F_2\delta^2 y_1 + E_4\delta^4 y_0 + F_4\delta^4 y_1 + \cdots \\
&= (1 - p)y_0 + E_2\delta^2 y_0 + E_4\delta^4 y_0 + \cdots \\
&\quad + (1 - q)y_1 + F_2\delta^2 y_1 + F_4\delta^4 y_1 + \cdots
\end{aligned}$$

where $p + q = 1$,

$$\begin{aligned}
E_2 &= -p(p - 1)(p - 2)/6 \\
F_2 &= -q(q - 1)(q - 2)/6 \\
E_4 &= -(p + 1)p(p - 1)(p - 2)(p - 3)/120 \\
F_4 &= -(q + 1)q(q - 1)(q - 2)(q - 3)/120
\end{aligned}$$

Throwback

The purpose of this device is to enable us to neglect the fourth difference term, when we have modified suitably the second difference term. We put

$$\delta_m^2 y_0 = \delta^2 y_0 - \lambda\delta^4 y_0, \qquad \delta_m^2 y_1 = \delta^2 y_1 - \lambda\delta^4 y_1$$

so that

$$\begin{aligned}
y_p &= y_0 + p\delta y_{\frac{1}{2}} + B_2(\delta_m^2 y_0 + \delta_m^2 y_1) + B_3\delta^3 y_{\frac{1}{2}} \\
&\quad + (B_4 + \lambda B_2)(\delta^4 y_0 + \delta^4 y_1) + \cdots
\end{aligned}$$

We now look for the best value to give to the constant λ. We have

$$\begin{aligned}
48(B_4 + \lambda B_2) &= (p^2 - p)(p^2 - p + 12\lambda - 2) \\
&= P(P + 2k) = P^2 + 2Pk
\end{aligned}$$

where $P = p^2 - p$, $k = 6\lambda - 1$.

If we now differentiate with respect to p and equate to 0, we find

$$(2P + 2k)(2p - 1) = 0$$

Either $p = \frac{1}{2}$, which gives $P(P + 2k)$ a maximum value equal to $(1 - 8k)/16$, or $P = -k$, which gives a minimum value equal to $-k^2$ at the two points given by $p^2 - p + k = 0$. We now choose k so that the maximum and minimum values are equal in magnitude, i.e.,

$$16k^2 = 1 - 8k, \qquad k = (-1 \pm \sqrt{2})/4$$

Since we require the two values of p which give the minimum value to

$48(B_4 + \lambda B_2)$ to lie between 0 and 1, it follows from the equation $p^2 - p + k = 0$ that k is positive.

Thus
$$k = (\sqrt{2} - 1)/4$$

and hence
$$\lambda = (k + 1)/6 = (3 + \sqrt{2})/24 = 0\!\cdot\!184.$$

With this value of λ, $\left|B_4 + \lambda B_2\right| < 0\!\cdot\!000224$ in $0 < p < 1$ and

$$\left|(B_4 + \lambda B_2)(\delta^4 y_0 + \delta^4 y_1)\right| < \tfrac{1}{2}\,\text{unit}$$

so long as the fourth differences are less than 1000 units. In this case we can safely truncate the Bessel formula after the third-order difference term.

The device of throwback can equally well be applied to the Everett formula, and if we assume that $\delta^2 y_0 = \delta^2 y_1$ and $\delta^4 y_0 = \delta^4 y_1$, we find the same value for λ.

EXAMPLES

(1) Calculate cosec $11° \, 8'$ using the values

$$\text{cosec } 11° \, 0' \ = 5\!\cdot\!2408, \qquad \text{cosec } 11° \, 6' \ = 5\!\cdot\!1942$$
$$\text{cosec } 11° \, 12' = 5\!\cdot\!1484, \qquad \text{cosec } 11° \, 18' = 5\!\cdot\!1034$$

Compare the result with the value obtained from four-figure tables using the mean difference column.

(2) Calculate tan $78° \, 52'$ using the values

$$\tan 78° \, 42' = 5\!\cdot\!0045, \qquad \tan 78° \, 48' = 5\!\cdot\!0504$$
$$\tan 78° \, 54' = 5\!\cdot\!0970, \qquad \tan 79° \, 0' \ = 5\!\cdot\!1446$$

(3) Given the table below of values of cot x, calculate cot $(0\!\cdot\!204)$ (*a*) by linear interpolation, (*b*) by quadratic interpolation.

x	0·2	0·21	0·22	0·23
cot x	4·9332	4·6917	4·4719	4·2709

(4) If values of x^4 between $x = 4\!\cdot\!0$ and $x = 4\!\cdot\!1$ are found by linear interpolation, find where the error is greatest. If the values of x^4 are required to the nearest integer, show that linear interpolation will not give the correct result for $x = 4\!\cdot\!07$.

(5) Obtain the formula

$$y_p = y_0 + p\nabla y_0 + \binom{p+1}{2}\nabla^2 y_0 + \binom{p+2}{3}\nabla^3 y_0 + \cdots$$

and use it to calculate $(4\!\cdot\!2)^4$ from a difference table for x^4, for $x = 0(1)4$.

(6) Find the maximum values of $|B_2|$, $|B_3|$, $|B_4|$, and $|B_5|$ in the range $0 < p < 1$. Deduce that if the moduli of $\delta^2 y$, $\delta^3 y$, $\delta^4 y$, and $\delta^5 y$ are less than 4, 60, 20, and 500 units respectively, then the corresponding terms in the Bessel interpolation formula are less than half a unit in magnitude.

(7) By putting $p = \frac{1}{2}$ in Bessel's formula obtain the relation

$$y_{\frac{1}{2}} = (-y_{-1} + 9y_0 + 9y_1 - y_2)/16$$

Use this result to evaluate the cube of 10·5.

(8) From the following table of values of $\sinh x$, calculate the value of $\sinh x$ for $x = 3\cdot42(0\cdot02)3\cdot58$.

x	3·0	3·2	3·4
$\sinh x$	10·018	12·246	14·965

x	3·6	3·8	4·0
$\sinh x$	18·285	22·339	27·290

(9) From the following table of values of $\cosh x$, calculate the values of $\cosh(1\cdot45)$, $\cosh(1\cdot5)$ and $\cosh(1\cdot55)$

x	1	1·2	1·4
$\cosh x$	1·54308	1·81066	2·15090

x	1·6	1·8	2·0
$\cosh x$	2·57746	3·10747	3·76220

(10) Given the values below of $y = \sin x$ for $x = 0(0\cdot5)2\cdot5$ show that

$$\delta^2 y_2 = -0\cdot2061, \quad \delta^2 y_3 = -0\cdot2442, \quad \delta_m^2 y_2 = -0\cdot2154, \quad \delta_m^2 y_3 = -0\cdot2552$$

Use the modified differences to calculate $\sin x$ for $x = 1\cdot1(0\cdot1)1\cdot4$.

$$y_0 = 0, \qquad y_1 = 0\cdot4794, \qquad y_2 = 0\cdot8415$$
$$y_3 = 0\cdot9975, \qquad y_4 = 0\cdot9093, \qquad y_5 = 0\cdot5985$$

(11) From four-figure and five-figure secant tables find the round-off error in the four-figure tables for the angles $68°(6')69°$. Draw up a difference table to show how these errors are propagated.

The Lagrange interpolation polynomial

We now make a fresh approach to the problem of interpolation. Let the values of y at $x = x_1, x_2, x_3$ be known. We can easily construct the quadratic expression in x that takes the same values at these points.
Put

$$L_1(x) = (x - x_2)(x - x_3)/(x_1 - x_2)(x_1 - x_3)$$
$$L_2(x) = (x - x_1)(x - x_3)/(x_2 - x_1)(x_2 - x_3)$$
$$L_3(x) = (x - x_1)(x - x_2)/(x_3 - x_1)(x_3 - x_2)$$

Each of these expressions is a quadratic in x that is zero for two of the three values of x and equals 1 at the other value of x. We now define $L(x)$ by the equation

$$L(x) = L_1(x)y_1 + L_2(x)y_2 + L_3(x)y_3$$

Then $L(x_1) = y_1, L(x_2) = y_2, L(x_3) = y_3$, so that $y = L(x)$ at $x = x_1, x_2, x_3$.
We can extend this to the case when n values of y are known. Put

$$L_i(x) = \frac{(x - x_1)(x - x_2)\ldots(x - x_{i-1})(x - x_{i+1})\ldots(x - x_n)}{(x_i - x_1)(x_i - x_2)\ldots(x_i - x_{i-1})(x_i - x_{i+1})\ldots(x_i - x_n)}$$

Then $L_i(x_i) = 1$, and $L_i(x_j) = 0$ if j is not equal to i, and so

$$y = L(x) = \sum_{i=1}^{n} L_i(x)y_i$$

at each of the points $x = x_1, x_2, \ldots, x_n$. Notice that these points need not be equally spaced.

$L(x)$ is known as the n-point Lagrange interpolation polynomial and we can express it in a slightly different form. We define the product $\pi(x)$ by the equation

$$\pi(x) = (x - x_1)(x - x_2)\ldots(x - x_n)$$

If we differentiate and then put $x = x_i$, we have

$$\pi'(x_i) = (x_i - x_1)(x_i - x_2)\ldots(x_i - x_{i-1})(x_i - x_{i+1})\ldots(x_i - x_n)$$

Hence
$$L_i(x) = \pi(x)/(x - x_i)\pi'(x_i)$$

and
$$L(x) = \sum_{i=1}^{n} \pi(x)y_i/(x - x_i)\pi'(x_i)$$

In the special case when y takes the same value k at each of the points $x_1, x_2, \ldots, x_n$ we have

$$L(x) = \sum_{i=1}^{n} L_i(x)k = k\sum_{i=1}^{n} L_i(x)$$

Then $L(x)$ will equal k at n points. But $L(x)$ is a polynomial of degree $(n - 1)$, since each $L_i(x)$ is of degree $(n - 1)$. The equation $L(x) = k$ can only have $(n - 1)$ solutions except when it is identically true. Thus we have proved that for all values of x

$$\sum_{i=1}^{n} L_i(x) = 1$$

This is the basis for an important check.

EXAMPLES

(1) If $y(1) = 12$, $y(2) = 15$, $y(5) = 25$ and $y(6) = 30$, find the four-point Lagrange interpolation polynomial that takes the same values as the function y at the given points.

$$
\begin{aligned}
L_1(x) &= (x - 2)(x - 5)(x - 6)/(-1)(-4)(-5) \\
       &= -(x^3 - 13x^2 + 52x - 60)/20 \\
L_2(x) &= (x - 1)(x - 5)(x - 6)/(1)(-3)(-4) \\
       &= +(x^3 - 12x^2 + 41x - 30)/12 \\
L_3(x) &= (x - 1)(x - 2)(x - 6)/(4)(3)(-1) \\
       &= -(x^3 - 9x^2 + 20x - 12)/12 \\
L_4(x) &= (x - 1)(x - 2)(x - 5)/(5)(4)(1) \\
       &= +(x^3 - 8x^2 + 17x - 10)/20
\end{aligned}
$$

We find that $L_1(x) + L_2(x) + L_3(x) + L_4(x) = 1$ as expected. The required polynomial is

$$
12L_1(x) + 15L_2(x) + 25L_3(x) + 30L_4(x)
$$
$$
= (4x^3 - 27x^2 + 233x + 510)/60
$$

We can verify that this takes the given values. The next example shows the lay-out commonly used.

(2) Given the values in the previous example, estimate the value of $y(4)$.

We first calculate the values of $|L_i(x)|$ at $x = 4$.

x_i	$|4 - x_i|$	$|$Numerator$|$	$|$Denominator$|$	$|L_i(4)|$
1	3	$2.1.2 = 4$	$1.4.5 = 20$	0·2
2	2	$3.1.2 = 6$	$1.3.4 = 12$	0·5
5	1	$3.2.2 = 12$	$4.3.1 = 12$	1·0
6	2	$3.2.1 = 6$	$5.4.1 = 20$	0·3

Now $L_2(4)$ is positive, since

$$
L_2(x) = \frac{(x - 1)(x - 5)(x - 6)}{(2 - 1)(2 - 5)(2 - 6)}
$$

In the same way $L_3(4)$ is positive, while $L_1(4)$ and $L_4(4)$ are negative. We can now insert the signs.

i	L_i	y_i	$L_i y_i$
1	-0.2	12	-2.4
2	$+0.5$	15	$+7.5$
3	$+1.0$	25	$+25.0$
4	-0.3	30	-9.0
Check	1.0		21.1

Estimated value of $y(4) = 21.1$.

(3) A polynomial y of the fourth degree in x takes the following values.

$$x \quad 0 \quad\quad 1 \quad\quad 2 \quad -1 \quad -2$$
$$y \quad 1 \quad -3 \quad -1 \quad\quad 2 \quad\quad 0$$

Without finding the polynomial, show that $y(3) = 25$.

(4) Calculate the value at $x = 5{\cdot}30$ of the Lagrange interpolation polynomial that takes these values:

$$x \quad\quad 5{\cdot}15 \quad\quad\quad 5{\cdot}48 \quad\quad\quad 5{\cdot}60 \quad\quad\quad 5{\cdot}99$$
$$y \quad 1{\cdot}63900 \quad 1{\cdot}70111 \quad 1{\cdot}72277 \quad 1{\cdot}79009$$

(5) Find the three-point Lagrange interpolation polynomial that takes the same value as the function $x^3 - 3x^2 + 3x$ at each of the points $x = 0, 1, 3$. Compare the values of the polynomial and the function at $x = 2$ and find the maximum difference between their values in the range $0 < x < 1$.

The error in Lagrange interpolation

The error is the difference between $y(x)$ and the Lagrange polynomial $L(x)$ at a given point. Let $y(x) = L(x)$ at the n points $x_1, x_2, \ldots, x_n$. At the point $x = \alpha$ the error will be $y(\alpha) - L(\alpha)$. We consider the function

$$y(x) - L(x) - \{y(\alpha) - L(\alpha)\}\pi(x)/\pi(\alpha)$$

where $\pi(x) = (x - x_1)(x - x_2)\ldots(x - x_n)$. This function is zero at the $(n + 1)$ points $x_1, x_2, \ldots, x_n, \alpha$. We assume that this function can be differentiated n times. Then by Rolle's theorem, its nth derivative vanishes at least once in any interval that contains the $(n + 1)$ zeros. Since the nth derivative of $L(x)$ is zero, we have

$$y^{(n)}(x) - \{y(\alpha) - L(\alpha)\}(n!)/\pi(\alpha) = 0$$

at some point, say $x = \xi$, so that

$$y(\alpha) - L(\alpha) = \pi(\alpha)y^{(n)}(\xi)/n!$$

If we know that the nth derivative is never greater than M in magnitude, we have

$$|\text{error}| < |\pi(\alpha)|M/n!$$

For example, if $y = \sin x$ we have $M = 1$, so that the error is less than $|\pi(\alpha)|/n!$.

The value of $\pi(\alpha)$ will of course vary with α and between two consecutive points at which $\pi(\alpha)$ is zero, $\pi(\alpha)$ will have a positive maximum or a negative minimum value. In the particular case when the points $x_1, x_2, \ldots, x_n$ are equally spaced the largest maxima of $|\pi(\alpha)|$ will occur when α lies between x_1 and x_2, and between x_{n-1} and x_n, while the smallest will occur in the middle of the range of values of x. We deduce that the error is more likely to be large at a point near one end of the range, and more likely to be small at a point near the middle of the range.

The weakness of Lagrange interpolation is the absence of the check provided by the difference table. If the method is to be used with confidence, we need some assurance that the function being evaluated behaves smoothly, i.e., that its derivatives do not take high values.

When the n points $x_1, x_2, \ldots, x_n$ are equally spaced, the expression above for the error becomes identical with the remainder term in the Newton forward difference formula, since we can write $(x - x_1)(x - x_2) \ldots (x - x_n)$ in the form $p(p - 1) \ldots (p - n + 1)h^n$. This is to be expected, since the polynomial of degree $(n - 1)$ that takes given values at n points is unique.

EXAMPLES

(1) Find the Lagrange polynomial $L(x)$ that takes the same values as x^4 at the points $x = \pm 1, \pm 2$. Draw the graph of the error, i.e., $x^4 - L(x)$.
(2) Sketch the graphs of

$$\pi(x) = (x - x_1)(x - x_2) \ldots (x - x_n)$$

for the cases $h = 0.25, 0.2, 0.1$ when $x_1 = 0$ and $x_n = 1$.
In each case find the maximum value of the modulus of $\pi(x)$ for values of x between 0 and 1.

Inverse interpolation

Our problem here is to calculate the value of x corresponding to a given value of y. We can do this by the Lagrange interpolation polynomial, treating y as the independent variable. The polynomial will be

$$L(y) = \sum_{i=1}^{n} L_i(y)x_i$$

and we calculate its value for the given value of y. However, if the values of x are equally spaced, it is usually preferable to make use of the difference table. We can employ Bessel's formula

$$y_p = y_0 + p\delta y_{\frac{1}{2}} + B_2(\delta^2 y_0 + \delta^2 y_1) + \cdots$$

in an iterative fashion, obtaining a sequence of values $p_1, p_2, p_3, \ldots$ converging to the required value of p.

We set out to find the value of x corresponding to the value of y_p, which lies between y_0 and y_1.

(a) First approximation

We neglect all the differences except the first order, and use

$$p_1 = (y_p - y_0)/\delta y_{\frac{1}{2}}$$

(b) Second approximation

We calculate B_2 for $p = p_1$, and find p_2 from

$$p_2 = \{y_p - y_0 - B_2(\delta^2 y_0 + \delta^2 y_1)\}/\delta y_{\frac{1}{2}}$$

(c) Third approximation

We calculate B_2 and B_3 for $p = p_2$, and find p_3 from

$$p_3 = \{y_p - y_0 - B_2(\delta^2 y_0 + \delta^2 y_1) - B_3 \delta^3 y_{\frac{1}{2}}\}/\delta y_{\frac{1}{2}}$$

We continue in this way until two successive approximations to p agree with one another.

The accuracy obtainable depends vitally on the number of significant figures in $\delta y_{\frac{1}{2}}$. The round-off error in the value of $\delta y_{\frac{1}{2}}$ may be close to one unit, so that if $\delta y_{\frac{1}{2}}$ equals n units the relative error in our final value of p may be close to $1/n$.

EXAMPLE

From the given values of $y = e^x$, solve the equation $e^x = 3 \cdot 206$:

x	1	1·1	1·2	1·3
y	2·718	3·004	3·320	3·669

$$
\begin{array}{llll}
x & y & & \\
1 & 2\cdot718 & & \\
 & & 286 & \\
1\cdot1 & 3\cdot004 & & 30 \\
 & & 316 & & 3 \\
1\cdot2 & 3\cdot320 & & 33 \\
 & & 349 & \\
1\cdot3 & 3\cdot669 & &
\end{array}
$$

We have $y_p = 3 \cdot 206$, $y_0 = 3 \cdot 004$, $\delta y_{\frac{1}{2}} = 0 \cdot 316$

$$p_1 = (3 \cdot 206 - 3 \cdot 004)/0 \cdot 316 = 0 \cdot 6 \quad \text{say}$$

There is no point in aiming at high accuracy at this stage.

$$B_2 = \tfrac{1}{4} p(p-1) = -0 \cdot 06 \quad \text{if } p = p_1$$
$$B_2(\delta^2 y_0 + \delta^2 y_1) = -0 \cdot 06(0 \cdot 030 + 0 \cdot 033) = -0 \cdot 004$$

Then
$$p_2 = (3 \cdot 206 - 3 \cdot 004 + 0 \cdot 004)/0 \cdot 316 = 0 \cdot 65$$
$$B_2 = \tfrac{1}{4} p(p-1) = -0 \cdot 057 \quad \text{if } p = p_2$$

so that
$$B_2(\delta^2 y_0 + \delta^2 y_1) = -0 \cdot 0036$$

$$B_3 = \tfrac{1}{6} p(p - \tfrac{1}{2})(p-1) = -0 \cdot 006 \quad \text{if } p = p_2$$

so that $B_3\delta^3 y_{\frac{1}{2}}$ is negligible. Then

$$p_3 = (3{\cdot}206 - 3{\cdot}004 + 0{\cdot}0036)/0{\cdot}316 = 0{\cdot}651$$

We then find $p_4 = 0{\cdot}651$, so that we have $p = 0{\cdot}651$, $x = 1{\cdot}1651$. However there is a possible error of $\frac{1}{2}$ unit in y_0 and a possible error of 1 unit in $\delta y_{\frac{1}{2}}$. In $(3{\cdot}206 - 3{\cdot}004)$ we have a possible error of about one part in 400 and in $0{\cdot}316$ a possible error of about 1 in 300. Since $1/400 + 1/300$ is approximately $1/170$, we have a possible error in $(3{\cdot}206 - 3{\cdot}004)/0{\cdot}316$ of about 1 in 170. The true value of p therefore lies between $0{\cdot}647$ and $0{\cdot}655$ and we are unable to locate it more closely. We should give our answer as $x = 1{\cdot}165$, and this will be correct to three decimal places.

An alternative method is based on the reversion of series. Let $(y - y_0)$ be expanded in a series of powers of $(x - x_0)$,

$$(y - y_0) = a(x - x_0) + b(x - x_0)^2 + \cdots$$

We assume that $(x - x_0)$ can be expanded in a series of powers of $(y - y_0)$,

$$(x - x_0) = A(y - y_0) + B(y - y_0)^2 + \cdots$$

We now substitute for $(x - x_0)$ in the first series.

$$(y - y_0) = aA(y - y_0) + (aB + bA^2)(y - y_0)^2 + \cdots$$

By equating coefficients we find $A = 1/a$, $B = -b/a^3$.

In the difference table above we have

$$y_0 = 3{\cdot}004, \qquad \Delta y_0 = 0{\cdot}316, \qquad \Delta^2 y_0 = 0{\cdot}033, \qquad h = 0{\cdot}1$$

Then

$$y = y_0 + (x - x_0)\Delta y_0/h + (x - x_0)(x - x_1)\Delta^2 y_0/2h^2 + \cdots$$
$$y - 3{\cdot}004 = (x - 1{\cdot}1)3{\cdot}16 + (x - 1{\cdot}1)(x - 1{\cdot}2)1{\cdot}65 + \cdots$$
$$= (x - 1{\cdot}1)2{\cdot}995 + (x - 1{\cdot}1)^2 1{\cdot}65 + \cdots$$
$$a = 2{\cdot}995, \qquad b = 1{\cdot}65, \qquad A = 0{\cdot}334, \qquad B = -0{\cdot}061$$
$$x - 1{\cdot}1 = 0{\cdot}334(3{\cdot}206 - 3{\cdot}004) - 0{\cdot}061(3{\cdot}206 - 3{\cdot}004)^2 + \cdots$$

Hence $x = 1{\cdot}1 + 0{\cdot}0650 = 1{\cdot}1650$.

This result suffers from the same uncertainty as before owing to round-off errors in Δy_0 and $\Delta^2 y_0$. The best that we can do is to give the answer as $x = 1{\cdot}165$, which is reliable to three decimal places.

Iterative interpolation

If we are given $(n + 1)$ points $P_0, P_1, \ldots, P_n$ on the graph of a single-valued function, we can find the equation $y = f(x)$ of the curve of degree $(n - 1)$ through the points $P_0, P_1, \ldots, P_{n-1}$, and also the equation $y = g(x)$ of the curve of degree $(n - 1)$ through the points $P_1, P_2, \ldots, P_n$. Then if $x = x_0$ at

P_0 and $x = x_n$ at P_n, the curve given by

$$y = \frac{(x - x_n)f(x) - (x - x_0)g(x)}{x_0 - x_n}$$

will pass through all the $(n + 1)$ points. For at $x = x_0$ this gives $y = f(x_0)$, at $x = x_n$ it gives $y = g(x_n)$, and at the other points we have $y = f(x) = g(x)$. The following two methods based on this idea are very suitable for use on a computer.

(a) Aitken's method

Suppose that four points $P_0(x_0, y_0)$, $P_1(x_1, y_1)$, $P_2(x_2, y_2)$, $P_3(x_3, y_3)$ are given. We begin by finding the equations of the three straight lines P_0P_1, P_0P_2, P_0P_3. Next we obtain the equations of parabolas through $P_0P_1P_2$ and $P_0P_1P_3$. Finally we obtain the equation of the cubic through the four points. An estimate of the value of y at some new value of x will be found by substituting this value of x in the cubic polynomial. In a computer program we introduce this value of x at the beginning of the calculation. We can set out the calculation as follows:

$$
\begin{array}{cccc}
x - x_0 & y_0 & & \\
 & & a_1 & \\
x - x_1 & y_1 & & b_1 \\
 & & a_2 & & c_1 \\
x - x_2 & y_2 & & b_2 \\
 & & a_3 & \\
x - x_3 & y_3 & &
\end{array}
$$

$$a_1 = \{(x - x_0)y_1 - (x - x_1)y_0\}/(x_1 - x_0)$$
$$a_2 = \{(x - x_0)y_2 - (x - x_2)y_0\}/(x_2 - x_0)$$
$$a_3 = \{(x - x_0)y_3 - (x - x_3)y_0\}/(x_3 - x_0)$$
$$b_1 = \{(x - x_1)a_2 - (x - x_2)a_1\}/(x_2 - x_1)$$
$$b_2 = \{(x - x_1)a_3 - (x - x_3)a_1\}/(x_3 - x_1)$$
$$c_1 = \{(x - x_2)b_2 - (x - x_3)b_1\}/(x_3 - x_2)$$

EXAMPLE

Let the given points be $(1, -7)$, $(3, 3)$, $(5, 5)$, $(6, -12)$ and let the value of y be required at $x = 4$.

$$
\begin{array}{cccc}
3 & -7 & & \\
 & & 8 & \\
1 & 3 & & 5 \\
 & & 2 & & 8 \\
-1 & 5 & & 2 \\
 & & -10 & \\
-2 & -12 & &
\end{array}
$$

Hence 8 is the value of the cubic at $x = 4$, so that our result is $y = 8$.

(b) Neville's method

In this scheme we take the equations of the straight lines through the points P_0P_1, P_1P_2, P_2P_3, and from them we find the equations of parabolas through the points $P_0P_1P_2$, $P_1P_2P_3$. The final cubic through all four points will of course be the same as in Aitken's method. We can use the same array, with the same values for a_1, b_1, and c_1.

$$
\begin{array}{cccc}
x - x_0 & y_0 & & \\
 & & a_1 & \\
x - x_1 & y_1 & & b_1 \\
 & & a_2 & & c_1 \\
x - x_2 & y_2 & & b_2 \\
 & & a_3 & \\
x - x_3 & y_3 & &
\end{array}
$$

$$a_1 = \{(x - x_0)y_1 - (x - x_1)y_0\}/(x_1 - x_0)$$
$$a_2 = \{(x - x_1)y_2 - (x - x_2)y_1\}/(x_2 - x_1)$$
$$a_3 = \{(x - x_2)y_3 - (x - x_3)y_2\}/(x_3 - x_2)$$
$$b_1 = \{(x - x_0)a_2 - (x - x_2)a_1\}/(x_2 - x_0)$$
$$b_2 = \{(x - x_1)a_3 - (x - x_3)a_2\}/(x_3 - x_1)$$
$$c_1 = \{(x - x_0)b_2 - (x - x_3)b_1\}/(x_3 - x_0)$$

For the same example as above we again find $y(4) = 8$ as shown below

$$
\begin{array}{cccc}
3 & -7 & & \\
 & & 8 & \\
1 & 3 & & 5 \\
 & & 4 & & 8 \\
-1 & 5 & & 10 \\
 & & 22 & \\
-2 & -12 & &
\end{array}
$$

Since the final polynomial will be identical with the Lagrange interpolation polynomial based on the same set of $(n + 1)$ points, the error in the interpolation by either of the above methods will be given by $\pi(x)y^{(n+1)}(\xi)/(n + 1)!$. When the number of given values is large, as in the process of interpolation in a table of values of a standard function, the convergence of the sequence of diagonal elements a_1, b_1, c_1, ... will indicate when no increase in accuracy will be obtained by including further values.

EXAMPLES

(1) From the given values of $\sin x°$, find the angle with sine equal to $0{\cdot}8$.

$$\sin 45° = 0{\cdot}7071, \qquad \sin 50° = 0{\cdot}7660$$
$$\sin 55° = 0{\cdot}8192, \qquad \sin 60° = 0{\cdot}8660$$

(2) Form a difference table for the values of e^x for $x = 0.2(0.01)0.25$ taken from a table giving five decimal places. Find the value of x for which $e^x = 1.25$.

(3) Form a difference table for $y = x^3$, $x = 0(1)4$, and use it to calculate the cube root of 10 correct to four decimal places.

(4) From the following table of values of cos x, calculate by inverse interpolation the value of π as accurately as the table allows.

x	1·0	1·2	1·4
cos x	0·54030	0·36236	0·16997

x	1·6	1·8	2·0
cos x	−0·02920	−0·22720	−0·41615

(5) The function $y(x)$ is a cubic polynomial in x taking the following exact values. Calculate y when $x = 5$, and calculate x when $y = 0$.

x	2·1	3·2	5·4	6·5
y	−2·4	−2·1	4·5	12·0

(6) Use (a) Aitken's method (b) Neville's method to calculate $y(3)$, given that $y(x)$ is a cubic polynomial in x which takes the following values.

x	−1	0	1	2
y	−1	3	1	−1

4

Numerical differentiation and integration

First-order derivatives

We define the differential operator D by defining Dy_n to be the value of dy/dx at $x = x_n$, and by defining $D^r y_n$ to be the value of the rth derivative of y at $x = x_n$. Note that the inverse operator D^{-1} must be used with caution, since $D^{-1}y$ is not unique. For example, $D^{-1}x^2$ equals $\frac{1}{3}x^3$ + constant.

If y can be expanded in a series of powers of x, say

$$y = a_0 + a_1 x + a_2 x^2 + \cdots + a_n x^n + \cdots$$

then by Taylor's theorem

$$a_0 = y(0) = y_0, \quad a_1 = y'(0) = Dy_0, \quad a_n = y^{(n)}(0)/n! = D^n y_0/n!$$

If we now put $x = h$, $y = y_1$, we have

$$y_1 = y_0 + hDy_0 + h^2 D^2 y_0/2! + \cdots + h^n D^n y_0/n! + \cdots$$

We can express this compactly in the form $y_1 = e^{hD}y_0$. Since $y_1 = Ey_0$ this gives the surprising relation $E = e^{hD}$. It must be remembered that the statement $Ey_0 = e^{hD}y_0$ is nothing more or less than the equation

$$y_1 = y_0 + hy_0' + h^2 y_0''/2! + \cdots + h^n y_0^{(n)}/n! + \cdots$$

If this Taylor series is convergent, then we are justified in using E and e^{hD} as equivalent operators. In particular, they will be equivalent when y is a polynomial.

From $E = e^{hD}$, we obtain $\log E = hD$ or

$$hD = \log\left(1 + \Delta\right) = \Delta - \tfrac{1}{2}\Delta^2 + \tfrac{1}{3}\Delta^3 - \cdots$$

Also since $E = 1/(1 - \nabla)$ we obtain

$$hD = -\log(1 - \nabla) = \nabla + \tfrac{1}{2}\nabla^2 + \tfrac{1}{3}\nabla^3 + \cdots$$

We can easily confirm the validity of the relation between D and Δ when it is applied to a polynomial by referring to the Newton formula,

$$y_p = y_0 + p\Delta y_0 + \binom{p}{2}\Delta^2 y_0 + \binom{p}{3}\Delta^3 y_0 + \cdots$$

which terminates when y is a polynomial in x. We differentiate with respect to p.

Since

$$\frac{\mathrm{d}}{\mathrm{d}p} y_p = \frac{\mathrm{d}}{\mathrm{d}p} y(x_0 + ph) = hy'(x_0 + ph) = hy'_p$$

we obtain

$$hy'_p = \Delta y_0 + (p - \tfrac{1}{2})\Delta^2 y_0 + (\tfrac{1}{2}p^2 - p + \tfrac{1}{3})\Delta^3 y_0 + \cdots$$

If we now put $p = 0$ we obtain

$$hy'_0 = \Delta y_0 - \tfrac{1}{2}\Delta^2 y_0 + \tfrac{1}{3}\Delta^3 y_0 - \cdots$$

Similarly by differentiating with respect to p the relation

$$y_p = y_0 + p\nabla y_0 + \binom{p+1}{2}\nabla^2 y_0 + \binom{p+2}{3}\nabla^3 y_0 + \cdots$$

we can obtain

$$hy'_0 = \nabla y_0 + \tfrac{1}{2}\nabla^2 y_0 + \tfrac{1}{3}\nabla^3 y_0 + \cdots$$

It follows that these formulae are valid when y is a polynomial in x.

In order to express the operator D in terms of central differences we put $E = e^{hD}$ in the relation $\delta = E^{\frac{1}{2}} - E^{-\frac{1}{2}}$. Then $\delta = 2 \sinh \tfrac{1}{2}hD$ and $hD = 2 \sinh^{-1} \tfrac{1}{2}\delta$. Now if $0 < x < 1$

$$\sinh^{-1} x = \int_0^x (1 + x^2)^{-\frac{1}{2}}\,\mathrm{d}x$$

$$= \int_0^x (1 - \tfrac{1}{2}x^2 + \tfrac{3}{8}x^4 - \cdots)\,\mathrm{d}x$$

$$= x - \tfrac{1}{6}x^3 + \tfrac{3}{40}x^5 - \cdots$$

Hence

$$hDy_{\frac{1}{2}} = 2(\tfrac{1}{2}\delta - \delta^3/48 + 3\delta^5/1280 - \cdots)y_{\frac{1}{2}}$$
$$= \delta y_{\frac{1}{2}} - \delta^3 y_{\frac{1}{2}}/24 + 3\delta^5 y_{\frac{1}{2}}/640 - \cdots$$

If we operate with hD on y_0 we obtain

$$hy'_0 = \delta y_0 - \delta^3 y_0/24 + 3\delta^5 y_0/640 - \cdots$$

and here we meet the difficulty that δy_0, $\delta^3 y_0$, $\delta^5 y_0$ are not entries in the difference table. Since $\mu^2 = 1 + \frac{1}{4}\delta^2$, we have $\mu(1 + \frac{1}{4}\delta^2)^{-\frac{1}{2}} = 1$. Hence we can write

$$hy'_0 = \mu(1 + \tfrac{1}{4}\delta^2)^{-\frac{1}{2}}(\delta - \delta^3/24 + 3\delta^5/640 - \cdots)y_0$$

We expand $(1 + \frac{1}{4}\delta^2)^{-\frac{1}{2}}$ to give $(1 - \frac{1}{8}\delta^2 + \frac{3}{128}\delta^4 - \cdots)$, and find

$$hy'_0 = \mu\delta y_0 - \tfrac{1}{6}\mu\delta^3 y_0 + \tfrac{1}{30}\mu\delta^5 y_0 - \cdots$$

where

$$\mu\delta y_0 = \tfrac{1}{2}(\delta y_{\frac{1}{2}} + \delta y_{-\frac{1}{2}}), \qquad \mu\delta^3 y_0 = \tfrac{1}{2}(\delta^3 y_{\frac{1}{2}} + \delta^3 y_{-\frac{1}{2}})$$

An alternative method of reaching these results would be to differentiate Bessel's formula with respect to p.

It is worthwhile to compare the two approximations

$$hy'_0 = \Delta y_0 - \tfrac{1}{2}\Delta^2 y_0 + \tfrac{1}{3}\Delta^3 y_0$$
$$hy'_0 = \mu\delta y_0 - \tfrac{1}{6}\mu\delta^3 y_0$$

The first can be written in terms of y_0, y_1, y_2, y_3 and we find

$$hy'_0 = (-11y_0 + 18y_1 - 9y_2 + 2y_3)/6$$

This has a maximum round-off error of $\{(2 + 9 + 18 + 11)/6\} \times \frac{1}{2}$ unit, i.e., $3\frac{1}{3}$ units, in the value of hy'_0. The second can be expressed in terms of y_2, y_1, y_{-1}, y_{-2}. We find

$$hy'_0 = (-y_2 + 8y_1 - 8y_{-1} + y_{-2})/12$$

with a maximum round-off error of $\{(1 + 8 + 8 + 1)/12\} \times \frac{1}{2}$ unit, i.e., $\frac{3}{4}$ unit, in the value of hy'_0. Thus the maximum round-off errors in the value of y'_0 are $10/3h$ and $3/4h$ units respectively, so that the second approximation is superior in this respect. It is also better with regard to truncation error. We take the first term omitted as a guide to the truncation error. In the first case it is $-\frac{1}{4}\Delta^4 y_0$ and in the second case it is $\frac{1}{30}\mu\delta^5 y_0$.

However, even with central difference formulae, the accuracy obtainable is often very limited owing to the effect of the size of the step-length. If we take h to be small, there is a loss of significant figures and a risk of large round-off error. With h large, the truncation error will be large unless more terms are included in the formula, but this will increase the round-off error. A balance must be struck between these conflicting tendencies. To illustrate this difficulty we use the forward difference formula to calculate the derivative of $y = \exp(-x)$ at $x = 2$. Its value to four decimal places is $-0\cdot1353$. Form a difference table with $h = 0\cdot01$.

x	y		
2	$0\cdot1353$		
		-13	
$2\cdot01$	$0\cdot1340$		0
		-13	
$2\cdot02$	$0\cdot1327$		

We find the value $-0.0013/h = -0.13$. We have only two significant figures in the result, the second of which is wrong.

Repeat with $h = 0.2$.

$$
\begin{array}{lll}
x & y & \\
2 & 0.1353 & \\
 & & -245 \\
2.2 & 0.1108 & \quad\quad 44 \\
 & & -201 \quad\quad\quad -7 \\
2.4 & 0.0907 & \quad\quad 37 \\
 & & -164 \\
2.6 & 0.0743 &
\end{array}
$$

We obtain the value

$$(-0.0245 - 0.0022 - 0.0002)/h = -0.1345$$

Here there is a round-off error in the first difference (-245) of possibly nearly 1 unit in size, in the second difference (44) of nearly 2 units, in the third difference (-7) of nearly 4 units. We conclude that the result lies between -0.1330 and -0.1360.

It is interesting to see what happens if we attempt to avoid these difficulties by using the Lagrange interpolation polynomial. Let the four-point polynomial based on the equally spaced points x_0, x_1, x_2, x_3 be

$$L(x) = \sum_{i=0}^{3} L_i(x)y_i$$

We differentiate with respect to x and then put $x = 0$. Since the coefficients of x in L_0, L_1, L_2, L_3 are $-11/6h, 3/h, -3/2h, 1/3h$ respectively we find

$$L'(0) = (-11y_0 + 18y_1 - 9y_2 + 2y_3)/6h$$

which we found earlier from the approximation

$$hy_0' = \Delta y_0 - \tfrac{1}{2}\Delta^2 y_0 + \tfrac{1}{3}\Delta^3 y_0$$

Again, take the five-point polynomial based on the equally spaced points $x_{-2}, x_{-1}, x_0, x_1, x_2$. We can write this in the form

$$L(x) = \pi(x)\left\{\frac{y_{-2}}{x + 2h} - \frac{4y_{-1}}{x + h} + \frac{6y_0}{x} - \frac{4y_1}{x - h} + \frac{y_2}{x - 2h}\right\}\Big/24h^4$$

where

$$\pi(x) = (x + 2h)(x + h)x(x - h)(x - 2h)$$

We differentiate and put $x = 0$, and obtain

$$L'(0) = (y_{-2} - 8y_{-1} + 8y_1 - y_2)/12h$$

(The middle term $y_0(x^2 - 4h^2)(x^2 - h^2)/4h^4$ has zero derivative at $x = 0$.)

This result is identical with that arrived at earlier from the approximation

$$hy'_0 = \mu\delta y_0 - \tfrac{1}{6}\mu\delta^3 y_0$$

This brings out the basic fact that when we use the difference table and assume that we can neglect differences of say the mth order, we are assuming that the function is a polynomial of degree $(m - 1)$, so that our results will agree with those derived from the $(m - 1)$-point Lagrange polynomial.

Second-order derivatives

From
$$hDy_0 = (\Delta - \tfrac{1}{2}\Delta^2 + \tfrac{1}{3}\Delta^3 - \cdots)y_0$$

we have
$$h^2 D^2 y_0 = (\Delta - \tfrac{1}{2}\Delta^2 + \tfrac{1}{3}\Delta^3 - \cdots)^2 y_0$$
$$= (\Delta^2 - \Delta^3 + \tfrac{11}{12}\Delta^4 - \cdots)y_0$$

Also from
$$hDy_0 = (\delta - \tfrac{1}{24}\delta^3 + \tfrac{3}{640}\delta^5 - \cdots)y_0$$
$$h^2 D^2 y_0 = (\delta - \tfrac{1}{24}\delta^3 + \tfrac{3}{640}\delta^5 - \cdots)^2 y_0$$

Hence
$$h^2 D^2 y_0 = \delta^2 y_0 - \tfrac{1}{12}\delta^4 y_0 + \tfrac{1}{90}\delta^6 y_0 - \cdots$$

EXAMPLES

(1) If we use the Lagrange type formula

$$L'(0) = (y_{-2} - 8y_{-1} + 8y_1 - y_2)/12h$$

show that the round-off error in the calculation of the derivative of e^{-x} at $x = 2$ with $h = 0.2$ will not exceed 4 units in the last decimal place.

(2) Calculate to four decimal places the derivative of $\sin x$ at $x = \pi/4$
   (a) using the values of $\sin 42°$, $\sin 44°$, $\sin 46°$, and $\sin 48°$ ($h = \pi/90$),
   (b) using the values of $\sin 30°$, $\sin 40°$, $\sin 50°$, and $\sin 60°$ ($h = \pi/18$).

Integration formulae

We first obtain formulae for the integral of y from x_0 to x_1 by integration with respect to p. We have

$$\int_{x_0}^{x_1} y \, dx = \int_0^1 y(x_0 + ph)h \, dp = h \int_0^1 y_p \, dp$$

From the forward difference formula

$$y_p = y_0 + p\Delta y_0 + \binom{p}{2}\Delta^2 y_0 + \binom{p}{3}\Delta^3 y_0 + \binom{p}{4}\Delta^4 y_0 + \cdots$$

we obtain by integrating with respect to p from 0 to 1

$$\frac{1}{h}\int_{x_0}^{x_1} y \, dx = y_0 + \tfrac{1}{2}\Delta y_0 - \tfrac{1}{12}\Delta^2 y_0 + \tfrac{1}{24}\Delta^3 y_0 - \tfrac{19}{720}\Delta^4 y_0 + \cdots$$

From the backward difference formulae

$$y_p = y_0 + p\nabla y_0 + \binom{p+1}{2}\nabla^2 y_0 + \binom{p+2}{3}\nabla^3 y_0 + \binom{p+3}{4}\nabla^4 y_0 + \cdots$$

$$= y_1 + (p-1)\nabla y_1 + \binom{p}{2}\nabla^2 y_1 + \binom{p+1}{3}\nabla^3 y_1 + \binom{p+2}{4}\nabla^4 y_1 + \cdots$$

we obtain in the same way

$$\frac{1}{h}\int_{x_0}^{x_1} y\,\mathrm{d}x = y_0 + \tfrac{1}{2}\nabla y_0 + \tfrac{5}{12}\nabla^2 y_0 + \tfrac{3}{8}\nabla^3 y_0 + \tfrac{251}{720}\nabla^4 y_0 + \cdots$$

$$= y_1 - \tfrac{1}{2}\nabla y_1 - \tfrac{1}{12}\nabla^2 y_1 - \tfrac{1}{24}\nabla^3 y_1 - \tfrac{19}{720}\nabla^4 y_1 - \cdots$$

These last two formulae are especially useful in the Adams–Bashforth method for the solution of differential equations.

To obtain a central difference integration formula for the integral of y from x_{-n} to x_n we write

$$\int_{-nh}^{nh} Y'(x)\,\mathrm{d}x = Y_n - Y_{-n} = (\mathrm{e}^{nhD} - \mathrm{e}^{-nhD})Y_0$$

We can expand this in powers of hD to give

$$2nh(1 + \tfrac{1}{6}n^2 h^2 D^2 + \tfrac{1}{120}n^4 h^4 D^4 + \cdots)DY_0$$

We now put $h^2 D^2 = \delta^2 - \delta^4/12 + \cdots$, $h^4 D^4 = \delta^4 - \cdots$, and $y(x) = Y'(x)$. This gives

$$\int_{-nh}^{nh} y(x)\,\mathrm{d}x = 2nh\left[1 + \tfrac{1}{6}n^2\delta^2 + \left(\frac{n^4}{120} - \frac{n^2}{72}\right)\delta^4 + \cdots\right]y_0$$

To obtain a central difference formula for the integral of y from x_0 to x_1 we integrate Bessel's formula

$$y_p = y_0 + p\delta y_{\frac{1}{2}} + B_2(\delta^2 y_0 + \delta^2 y_1) + B_3\delta^3 y_{\frac{1}{2}} + \cdots$$

We obtain

$$\frac{1}{h}\int_{x_0}^{x_1} y\,\mathrm{d}x = y_0 + \tfrac{1}{2}\delta y_{\frac{1}{2}} - \tfrac{1}{24}(\delta^2 y_0 + \delta^2 y_1) + \tfrac{11}{1440}(\delta^4 y_0 + \delta^4 y_1) + \cdots$$

$$= \tfrac{1}{2}(y_0 + y_1) - \tfrac{1}{12}\mu\delta^2 y_{\frac{1}{2}} + \tfrac{11}{720}\mu\delta^4 y_{\frac{1}{2}} - \cdots$$

$$= \tfrac{1}{2}(y_0 + y_1) - \tfrac{1}{12}\mu(\delta y_1 - \delta y_0) + \tfrac{11}{720}\mu(\delta^3 y_1 - \delta^3 y_0) - \cdots$$

We can immediately obtain from this an expression for the integral of y from x_0 to x_n. We have

$$\frac{1}{h}\int_{x_1}^{x_2} y\,\mathrm{d}x = \tfrac{1}{2}(y_1 + y_2) - \tfrac{1}{12}\mu(\delta y_2 - \delta y_1) + \tfrac{11}{720}\mu(\delta^3 y_2 - \delta^3 y_1) - \cdots$$

and similar formulae for the intervals $(x_2, x_3), \ldots, (x_{n-1}, x_n)$. If we add these formulae together we obtain

$$\int_{x_0}^{x_n} y \, dx = \tfrac{1}{2}h(y_0 + 2y_1 + 2y_2 + \cdots + 2y_{n-1} + y_n)$$
$$-\frac{h}{12}\mu(\delta y_n - \delta y_0) + \frac{11h}{720}\mu(\delta^3 y_n - \delta^3 y_0) - \cdots$$

This is known as the Gauss–Encke formula. Notice that the expression

$$\tfrac{1}{2}h(y_0 + 2y_1 + 2y_2 + \cdots + 2y_{n-1} + y_n)$$

is the value given for the integral by the trapezium rule. We can regard the first difference term as the first correction term, and the next term as the second correction term. The Gauss–Encke formula converges rapidly, but for its use we require to know values of y at points outside the range of integration. (To find $\mu\delta y_0$ for instance we need to know the value of y_{-1} so that we can find $\delta y_{-\frac{1}{2}}$.)

To overcome this disadvantage we express $\mu\delta$ in terms of Δ and ∇.

$$\mu\delta y_0 = \tfrac{1}{2}(E - E^{-1})y_0 = \tfrac{1}{2}(1 + \Delta)y_0 - \tfrac{1}{2}(1 + \Delta)^{-1}y_0$$
$$= (\Delta - \tfrac{1}{2}\Delta^2 + \tfrac{1}{2}\Delta^3 - \tfrac{1}{2}\Delta^4 + \cdots)y_0$$
$$\mu\delta^3 y_0 = \mu\delta(\delta^2)y_0 = \Delta^3 y_0 - \tfrac{3}{2}\Delta^4 y_0 + \cdots$$

since $\delta^2 = \Delta^2/(1 + \Delta)$.

Also since $\delta^2 = \nabla^2/(1 - \nabla)$ we have

$$\mu\delta y_n = \tfrac{1}{2}(1 - \nabla)^{-1}y_n - \tfrac{1}{2}(1 - \nabla)y_n$$
$$= (\nabla + \tfrac{1}{2}\nabla^2 + \tfrac{1}{2}\nabla^3 + \tfrac{1}{2}\nabla^4)y_n$$
$$\mu\delta^3 y_n = \nabla^3 y_n + \tfrac{3}{2}\nabla^4 y_n + \cdots$$

If we use these relations in the Gauss–Encke formula we obtain

$$\int_{x_0}^{x_n} y \, dx = \tfrac{1}{2}h(y_0 + 2y_1 + 2y_2 + \cdots + 2y_{n-1} + y_n)$$
$$+\frac{h}{12}(\Delta y_0 - \nabla y_n) - \frac{h}{24}(\Delta^2 y_0 + \nabla^2 y_n) + \frac{19h}{720}(\Delta^3 y_0 - \nabla^3 y_n)$$
$$-\frac{3h}{160}(\Delta^4 y_0 + \nabla^4 y_n) + \cdots$$

This is Gregory's formula, which is used when the value of y is not known outside the range of integration.

EXAMPLE

Evaluate $\int_0^1 \cos x \, dx$ as accurately as five-figure tables will allow, using a step-length $h = 0.2$.

x	$\cos x$					
-0.4	0.92106					
		5901				
-0.2	0.98007		-3908			
		1993		-78		
0	1.00000		-3986		156	
		-1993		78		3
0.2	0.98007		-3908		159	
		-5901		237		-16
0.4	0.92106		-3671		143	
		-9572		380		-10
0.6	0.82534		-3291		133	
		-12863		513		-21
0.8	0.69671		-2778		112	
		-15641		625		-29
1.0	0.54030		-2153		83	
		-17794		708		
1.2	0.36236		-1445			
		-19239				
1.4	0.16997					

(*a*) The trapezium rule

$$\tfrac{1}{2}(1.0000 + 0.54030) = 0.77015$$
$$0.98007 + 0.92106 + 0.82534 + 0.69671 = \underline{3.42318}$$
$$4.19333$$

Since $h = 0.2$, this gives $\underline{0.83867}$.

(*b*) The Gauss–Encke formula

$$\mu\delta y_0 = 0, \qquad \mu\delta y_5 = \tfrac{1}{2}(-0.17794 - 0.15641)$$
$$\mu\delta^3 y_0 = 0, \qquad \mu\delta^3 y_5 = \tfrac{1}{2}(0.00708 + 0.00625)$$

Hence
$$-\tfrac{1}{12}\mu(\delta y_5 - \delta y_0) = 0.01393(1)$$
$$\tfrac{11}{720}\mu(\delta^3 y_5 - \delta^3 y_0) = 0.00010(2)$$
$$\text{Trapezium term} = \underline{4.19333}$$
$$4.20736(3)$$

We now multiply by h and obtain the result $\underline{0.84147}$.

(*c*) Gregory's formula

For this we use only the part of the difference table based on the values of $\cos x$ from $x = 0$ to $x = 1$.

$$
\begin{aligned}
\text{Trapezium term} &= & 4\cdot19333 \\
\tfrac{1}{12}(\Delta y_0 - \nabla y_n) = \tfrac{1}{12}(-0\cdot01993 + 0\cdot15641) &= & 0\cdot01137 \ (3) \\
-\tfrac{1}{24}(\Delta^2 y_0 + \nabla^2 y_n) = \tfrac{1}{24}(0\cdot03908 + 0\cdot02778) &= & 0\cdot00278 \ (6) \\
\tfrac{19}{720}(\Delta^3 y_0 - \nabla^3 y_n) = \tfrac{19}{720}(0\cdot00237 - 0\cdot00513) &= & -0\cdot00007 \ (3) \\
-\tfrac{3}{160}(\Delta^4 y_0 + \nabla^4 y_n) = -\tfrac{3}{160}(0\cdot00143 + 0\cdot00133) &= & -0\cdot00005 \ (2) \\
\hline
&& 4\cdot20736 \ (4)
\end{aligned}
$$

Again since $h = 0\cdot2$, the result is $\underline{0\cdot84147}$.

The maximum error in the trapezium term due to round-off in the values of $\cos x$ is $2\tfrac{1}{4}$ units. Since this is multiplied by h, this could give rise to an error of nearly $\tfrac{1}{2}$ unit in the final result. We cannot therefore be certain that the last digit in our result is correct.

When y is a polynomial of degree n in x, the Gauss–Encke formula and the Gregory formula both terminate, since the differences of order $(n + 1)$ will be zero. For $n = 2$, the Gregory formula gives

$$
\begin{aligned}
\frac{1}{h}\int_{x_0}^{x_2} y \, dx &= \tfrac{1}{2}y_0 + y_1 + \tfrac{1}{2}y_2 + \tfrac{1}{12}(\Delta y_0 - \nabla y_2) - \tfrac{1}{24}(\Delta^2 y_0 + \nabla^2 y_2) \\
&= \tfrac{1}{2}y_0 + y_1 + \tfrac{1}{2}y_2 + \tfrac{1}{12}(y_1 - y_0 - y_2 + y_1) - \tfrac{1}{12}(y_2 - 2y_1 + y_0)
\end{aligned}
$$

This gives Simpson's rule

$$
\int_{x_0}^{x_2} y \, dx = \tfrac{1}{3}h(y_0 + 4y_1 + y_2)
$$

For $n = 3$, we find in the same way the three-eighths rule

$$
\int_{x_0}^{x_3} y \, dx = \tfrac{3}{8}h(y_0 + 3y_1 + 3y_2 + y_3)
$$

For $n = 4$, we find Boole's rule

$$
\int_{x_0}^{x_4} y \, dx = \frac{2h}{45}(7y_0 + 32y_1 + 12y_2 + 32y_3 + 7y_4)
$$

We shall see that these are exact when y is a polynomial of degree 3, 3, and 5 respectively. We can use them as approximations if we can assume that y differs by a negligible amount from a polynomial of the required degree. They can be used without a difference table, since the result is expressed directly in terms of the ordinates. This is characteristic of Lagrangian formulae, which we consider next.

EXAMPLES

(1) Show that

$$\int_{x_{-1}}^{x_1} y \, dx = 2h(1 + \tfrac{1}{6}\delta^2 - \tfrac{1}{180}\delta^4 + \cdots)y_0$$

(2) Obtain the relations

$$\int_{x_0}^{x_2} y \, dx = h(2 + \tfrac{1}{3}\nabla^2 + \tfrac{1}{3}\nabla^3 + \tfrac{29}{90}\nabla^4 + \cdots)y_1$$
$$= h(2 - 2\nabla + \tfrac{1}{3}\nabla^2 - \tfrac{1}{90}\nabla^4 + \cdots)y_2$$

(3) Verify that the coefficients of Δy_0, $\Delta^2 y_0$, $\Delta^3 y_0$ in Gregory's formula are given by

$$-\int_0^1 h\binom{p}{2}dp, \qquad -\int_0^1 h\binom{p}{3}dp, \qquad -\int_0^1 h\binom{p}{4}dp$$

(4) Calculate $\int_0^1 \sin x \, dx$ with $h = 0.2$ by the Gauss–Encke formula and by Gregory's formula, using values of $\sin x$ from five-figure tables.

(5) Evaluate $\int \sec x° \, dx$ from $x = 0$ to $x = 45$ using $h = 5$, working with four decimal places. Compare the answer with the value obtained from $\int \sec x \, dx = \log(\sec x + \tan x)$.

(6) Calculate the integral of $\exp(-x^2)$ from $x = -0.2$ to $x = +0.2$ using the following table of values:

x	0	0·1	0·2	0·3	0·4
$\exp(-x^2)$	1	0·9900	0·9608	0·9139	0·8521

(7) Evaluate the integral of $J_0(x)$ over the range $0 \leqslant x \leqslant 0.5$ using the following table of values of the Bessel function $J_0(x)$ and its differences.

x	$J_0(x)$	δ^2	δ^4
0	1·00000	−500	4
0·1	0·99750	−498	5
0·2	0·99002	−491	0
0·3	0·97763	−484	7
0·4	0·96040	−470	2
0·5	0·93847	−454	5

(8) Evaluate the integral of $\exp(1/x)$ from $x = 1$ to $x = 2$ using (i) a step-length $h = 0.25$, (ii) a step-length $h = 0.2$.

(9) Establish the formula

$$\int_{-5h}^{5h} y \, dx = 10h(y_0 + 25\delta^2 y_0/6 + 175\delta^4 y_0/36 + \cdots)$$

Use the difference table above for $\cos x$ to evaluate the integral of $\cos x$ from -1 to $+1$.

(10) From Gregory's formula deduce the result

$$\int_0^{nh} y\, dx = h(\tfrac{1}{2}y_0 + y_1 + y_2 + \cdots + y_{n-1} + \tfrac{1}{2}y_n)$$
$$-(h^2/12)(y_n' - y_0') + (h^4/720)(y_n''' - y_0''') - \cdots$$

By putting $y = x^3$, $h = 1$ use this result to find the sum of the cubes of the first n integers.

Newton–Cotes integration formulae

The three-point Lagrange interpolation polynomial based on the points $x = -h, 0, h$ is

$$\{x(x - h)y_{-1} - 2(x^2 - h^2)y_0 + x(x + h)y_1\}/2h^2$$

If we integrate this polynomial from $x = -h$ to $x = h$ we obtain $\tfrac{1}{3}h(y_{-1} + 4y_0 + y_1)$. If we ignore the difference in value between y and the Lagrange polynomial, this gives Simpson's rule

$$\int_{-h}^{h} y\, dx = \tfrac{1}{3}h(y_{-1} + 4y_0 + y_1)$$

We can write this in the form

$$\int_{x_0}^{x_2} y\, dx = \tfrac{1}{3}h(y_0 + 4y_1 + y_2)$$

and hence we obtain

$$\int_{x_0}^{x_4} y\, dx = \tfrac{1}{3}h(y_0 + 4y_1 + 2y_2 + 4y_3 + y_4)$$
$$\int_{x_0}^{x_6} y\, dx = \tfrac{1}{3}h(y_0 + 4y_1 + 2y_2 + 4y_3 + 2y_4 + 4y_5 + y_6)$$

In this form the formula is known as the parabolic rule. We can apply the rule to any range of integration by dividing the range into an even number of sub-intervals each of length h. It is widely used, partly because the coefficients are simple and partly because the error is usually small. The difference between the function y and the three-point Lagrange polynomial above is

$$(x + h)x(x - h)y^{(3)}(\xi)/3!$$

where ξ lies between $-h$ and h. It follows that the error in Simpson's rule is equal to

$$\int_{-h}^{h} (x + h)x(x - h)y^{(3)}(\xi)\, dx/3!$$

Since ξ varies with x, this expression is not easy to simplify.

Now if y is a quadratic in x, it equals the three-point Lagrange polynomial, so that the integration formula is exact. If y is a cubic in x, $y^{(3)}(\xi)$ will be a constant. We can then evaluate the expression for the error and since

$$\int_{-h}^{h} (x + h)x(x - h)\, dx = 0$$

we find that Simpson's rule is exact not only for a quadratic but also for a cubic. This unexpected advantage is derived from the fact that by symmetry the integral of x^3 from $-h$ to h is zero.

To find the error in Simpson's rule we assume that y can be differentiated four times with respect to x in the range $(-h, h)$. Let $F(x) = \int y\, dx$, so that

$$\int_{-h}^{h} y\, dx = F(h) - F(-h)$$

and

$$\frac{d}{dh} \int_{-h}^{h} y\, dx = F'(h) + F'(-h) = y(h) + y(-h)$$

Consider the expression

$$F(h) - F(-h) - \tfrac{1}{3}h\{y(-h) + 4y(0) + y(h)\}$$

as a function of h. Its first derivative is

$$y(h) + y(-h) - \tfrac{1}{3}\{y(-h) + 4y(0) + y(h)\} - \tfrac{1}{3}h\{-y'(-h) + y'(h)\}$$

which is zero at $h = 0$. We find that the second derivative also vanishes at $h = 0$, while the third derivative is

$$-\tfrac{1}{3}h\{-y'''(-h) + y'''(h)\}$$

Then by the mean value theorem we have

$$[F(h) - F(-h) - \tfrac{1}{3}h\{y(-h) + 4y(0) + y(h)\}]/h^5$$
$$= -\tfrac{1}{3}\{-y'''(-\xi_1) + y'''(\xi_1)\}/60\xi_1$$
$$= -\tfrac{2}{3}\xi_1\,\{y^{(4)}(\xi)\}/60\xi_1 = -\tfrac{1}{90}y^{(4)}(\xi)$$

where $0 < \xi_1 < h$ and $-\xi_1 < \xi < \xi_1$.

Thus we have shown that the error in Simpson's rule equals $-\tfrac{1}{90}h^5 y^{(4)}(\xi)$.

By integrating the four-point Lagrange interpolation formula based on the points $x = 0, h, 2h, 3h$ we can obtain the three-eighths rule

$$\int_{x_0}^{x_3} y\, dx = \tfrac{3}{8}h(y_0 + 3y_1 + 3y_2 + y_3) - \tfrac{3}{80}h^5 y^{(4)}(\xi)$$

where $0 < \xi < x_3$. The error term here has a larger coefficient than that in Simpson's rule. However, if these formulae are used to evaluate the integral of y over an interval (a, b), in one case $h = (b - a)/2$ and in the other $h = (b - a)/3$, so that in fact the coefficient of the fourth derivative is smaller in the three-eighths rule.

To obtain the error term we revert to the expression in terms of divided differences for the error in the Lagrange interpolation formula. In this case it is

$$x(x - h)(x - 2h)(x - 3h)[0, h, 2h, 3h, x]$$

which is equal to

$$x(x - h)(x - 2h)\{[0, h, 2h, x] - [0, h, 2h, 3h]\}$$

Integrate by parts, using $x^2(x - 2h)^2/4$ as the integral of $x(x - h)(x - 2h)$. Since

$$\tfrac{1}{4}x^2(x - 2h)^2\{[0, h, 2h, x] - [0, h, 2h, 3h]\} = 0 \quad \text{at } x = 0 \text{ or } 3h$$

we have

$$-\int_0^{3h} \tfrac{1}{4}x^2(x - 2h)^2 \frac{\mathrm{d}}{\mathrm{d}x}[0, h, 2h, x]\,\mathrm{d}x$$

$$= -\int_0^{3h} \tfrac{1}{4}x^2(x - 2h)^2[0, h, 2h, x, x]\,\mathrm{d}x$$

$$= -[0, h, 2h, \xi, \xi]\int_0^{3h} \tfrac{1}{4}x^2(x - 2h)^2\,\mathrm{d}x = -(\tfrac{9}{10})h^5[0, h, 2h, \xi, \xi]$$

We have been able to use the mean value theorem here since $x^2(x - 2h)^2$ is always positive. We now replace $[0, h, 2h, \xi, \xi]$ by $y^{(4)}(\xi_1)/4!$ so that the error is given by $-3h^5 y^{(4)}(\xi_1)/80$.

By integrating the four-point Lagrange interpolation formula based on the points $x = 0, h, 2h, 3h$ over the interval $(0, h)$, we obtain the following useful result

$$\int_0^h y\,\mathrm{d}x = \tfrac{1}{24}h\{9y(0) + 19y(h) - 5y(2h) + y(3h)\}$$

with an error term $-\tfrac{19}{720}h^5 y^{(4)}(\xi)$, $0 < \xi < 3h$.

Simpson's rule and the three-eighths rule are examples of closed-type formulae, in which the range of integration does not extend outside the interval containing the points on which the Lagrange polynomial is based. In the step-by-step solution of differential equations, we shall need open-type formulae in which the range of integration extends outside this interval. As is to be expected, the error term in an open-type formula is larger than that in the closed-type formula using the same number of points. The most useful of the open-type formulae is

$$\int_{-2h}^{2h} y\,\mathrm{d}x = \tfrac{4}{3}h(2y_{-1} - y_0 + 2y_1) + \tfrac{14}{45}h^5 y^{(4)}(\xi_1)$$

where $x_{-2} < \xi_1 < x_2$, known as Milne's formula. If we integrate the Lagrange polynomial

$$\{x(x - h)y_{-1} - 2(x^2 - h^2)y_0 + x(x + h)y_1\}/2h^2$$

between $-2h$ and $2h$ we obtain $\frac{4}{3}h(2y_{-1} - y_0 + 2y_1)$. The error term is given by

$$\int_{-2h}^{2h} (x + h)x(x - h)[-h, 0, h, x]\,\mathrm{d}x$$

$$= \int_{-2h}^{2h} \frac{1}{4}\frac{\mathrm{d}}{\mathrm{d}x}(x^4 - 2x^2h^2 - 8h^4)[-h, 0, h, x]\,\mathrm{d}x$$

$$= \int_{-2h}^{2h} \tfrac{1}{4}(x^4 - 2x^2h^2 - 8h^4)[-h, 0, h, x, x]\,\mathrm{d}x$$

$$= \tfrac{1}{4}[-h, 0, h, \xi, \xi]\int_{-2h}^{2h} (x^4 - 2x^2h^2 - 8h^4)\,\mathrm{d}x$$

$$= [-h, 0, h, \xi, \xi](112h^5/15) = \tfrac{14}{45}h^5 y^{(4)}(\xi_1)$$

It must be emphasized that an approximate formula is of little use unless we possess an estimate of the magnitude of the error term.

When such an estimate is not available, we can employ Romberg's method of applying the trapezium rule. The first correction to the trapezium rule is shown by the Gauss–Encke formula to be $-\frac{1}{12}h\mu(\delta y_n - \delta y_0)$. This is approximately equal to $-\frac{1}{12}h^2(y'_n - y'_0)$, so that the truncation error in the trapezium rule is of the order of h^2. If we evaluate an integral over $[a, b]$ by the trapezium rule using (a) $h = (b - a)/2^n$, (b) $h = (b - a)/2^{n+1}$, the error in the first result T_n will be approximately four times that in the second result T_{n+1}, since the step-lengths are in the ratio $2:1$. Hence a better approximation to the correct result is given by

$$U_n = T_{n+1} + \tfrac{1}{3}(T_{n+1} - T_n) = (4T_{n+1} - T_n)/3$$

In the case $n = 2$, with $a = 0$ and $b = 8h$, we have

$$T_2 = 2h(\tfrac{1}{2}y_0 + y_2 + y_4 + y_6 + \tfrac{1}{2}y_8)$$
$$T_3 = h(\tfrac{1}{2}y_0 + y_1 + \cdots + y_7 + \tfrac{1}{2}y_8)$$
$$U_2 = (4T_3 - T_2)/3 = \tfrac{1}{3}h(y_0 + 4y_1 + 2y_2 + \cdots + 4y_7 + y_8)$$

This is the parabolic rule (with eight sub-intervals), in which the error is of the order of h^4. We can easily show that U_n is the result given by the parabolic formula with 2^{n+1} sub-intervals. We now calculate T_{n+2}, using the trapezium rule with $h = (b - a)/2^{n+2}$ and find the value of

$$U_{n+1} = T_{n+2} + \tfrac{1}{3}(T_{n+2} - T_{n+1})$$

The error in U_n will be approximately 2^4 times that in U_{n+1} so that a better approximation for the integral is

$$V_n = U_{n+1} + (U_{n+1} - U_n)/15$$

The truncation error in V_n will be of the order of h^6 where $h = (b - a)/2^n$.

We next find T_{n+3}, U_{n+2}, and V_{n+1}, and if V_{n+1} differs from V_n we form $V_{n+1} + (V_{n+1} - V_n)/63$.

EXAMPLES

(1) Obtain the closed-type formula

$$\int_0^{4h} y\,dx = \tfrac{2}{45}h(7y_0 + 32y_1 + 12y_2 + 32y_3 + 7y_4)$$

Verify that it is exact for $y = x^5$ and find the error when it is applied to $y = x^6$, with $h = 0 \cdot 5$.

(2) Obtain the open-type formula

$$\int_0^{3h} y\,dx = \tfrac{3}{2}h(y_1 + y_2) + \tfrac{3}{4}h^3 y''(\xi)$$

by dividing the range of integration into the two parts $(0, 2h)$ and $(2h, 3h)$.

(3) By integrating the four-point Lagrange interpolation polynomial based on the points $x = h, 2h, 3h, 4h$ obtain the open-type formula

$$\int_0^{5h} y\,dx = \tfrac{5}{24}h(11y_1 + y_2 + y_3 + 11y_4)$$

and show that it is exact if y is a polynomial of degree 3 or less.

(4) By expressing $\mu\delta$ and $\mu\delta^3$ in terms of hD, obtain the formula

$$\int_{x_0}^{x_1} y\,dx = \tfrac{1}{2}h(y_0 + y_1) - \frac{h^2}{12}(y_1' - y_0') + \frac{h^4}{720}(y_1''' - y_0''') \ldots$$

from the Gauss–Encke formula.

(5) Obtain the formula

$$\int_0^h y\,dx = \tfrac{1}{24}h\{9y(0) + 19y(h) - 5y(2h) + y(3h)\}$$

and find the error when it is applied to $y = x^4$.

Chebyshev integration formulae

We now consider a set of integration formulae which are characterized by the fact that the coefficients of the ordinates in the relation are all equal. The three-point formula is of the form

$$\int_{-1}^{1} y\,dx = k(y_1 + y_2 + y_3)$$

We can find a constant k and three points in the range $(-1, 1)$ such that this formula is exact for any cubic polynomial in x. Put y equal to a constant.

This shows that $k = \frac{2}{3}$. Then we require the formula to be exact for $y = x$, x^2, x^3, so that

$$x_1 + x_2 + x_3 = 0, \qquad x_1^2 + x_2^2 + x_3^2 = 1, \qquad x_1^3 + x_2^3 + x_3^3 = 0$$

These equations are satisfied by

$$x_1 = -1/\sqrt{2}, \qquad x_2 = 0, \qquad x_3 = 1/\sqrt{2}$$

Thus

$$\int_{-1}^{1} y \, dx = \tfrac{2}{3}\{y(-1/\sqrt{2}) + y(0) + y(1/\sqrt{2})\}$$

Note that the three values of y employed are given equal weight, so that a large round-off error in one particular value does not have undue influence on the result.

The error term in the formula can be expressed as

$$\int_{-1}^{1} (x^3 - \tfrac{1}{2}x)[x_1, x_2, x_3, x] \, dx$$

$$= \left[\tfrac{1}{4}(x^4 - x^2)[x_1, x_2, x_3, x]\right]_{-1}^{1} - \int_{-1}^{1} \tfrac{1}{4}x^2(x^2 - 1)[x_1, x_2, x_3, x, x] \, dx$$

$$= \tfrac{1}{15}[x_1, x_2, x_3, \xi, \xi] = y^{(4)}(\xi_1)/360 \quad \text{where} \; -1 < \xi_1 < 1$$

This shows that the formula is surprisingly accurate. Its disadvantage is that we have to calculate the value of y at $x = \pm 1/\sqrt{2}$.

The four-point Chebyshev formula is

$$\int_{-1}^{1} y \, dx = \tfrac{1}{2}(y_1 + y_2 + y_3 + y_4)$$

where the constant factor $\frac{1}{2}$ is chosen to make the relation hold when y is constant. If we can choose four points such that

$$x_1 + x_2 + x_3 + x_4 = 0, \qquad x_1^2 + x_2^2 + x_3^2 + x_4^2 = \tfrac{4}{3}$$
$$x_1^3 + x_2^3 + x_3^3 + x_4^3 = 0, \qquad x_1^4 + x_2^4 + x_3^4 + x_4^4 = \tfrac{4}{5}$$

the formula will be exact for $y = x^n$, $n = 0, 1, 2, 3, 4$.

It is reasonable to assume that by symmetry $x_1 = -x_4$, $x_2 = -x_3$, so that $x_1^2 + x_2^2 = \frac{2}{3}$, $x_1^4 + x_2^4 = \frac{2}{5}$. These equations are satisfied by $x_1^2 = (5 + 2\sqrt{5})/15$, $x_2^2 = (5 - 2\sqrt{5})/15$, and hence we find that the four points required are

$$x = \pm 0\cdot 1876, \qquad x = \pm 0\cdot 7947$$

Since we have to calculate the values of y at these points, this formula is more suitable for use on a computer than on a desk-calculator. By the method used above, the error term can be shown to be given by $2y^{(6)}(\xi)/42{,}525$.

In the same way, we can obtain formulae of this type for 5, 6, 7, and 9 points. If the number of points is 8 or 10 or more, the set of equations to be solved has two or more complex roots.

Gaussian integration formulae

Here again we obtain formulae of great accuracy by choosing the values of x employed. We take points $x_1, x_2, \ldots, x_n$ such that the polynomial

$$\pi(x) = (x - x_1)(x - x_2) \ldots (x - x_n)$$

satisfies the relations

$$\int_{-1}^{1} \pi(x)x^r \, dx = 0, \quad r = 0, 1, 2, \ldots, n - 1$$

We then arrive at a formula of the form

$$\int_{-1}^{1} y \, dx = A_1 y_1 + A_2 y_2 + \cdots + A_n y_n$$

In the more general case, a positive weighting function $w(x)$ is introduced and we have

$$\int_{-1}^{1} w(x)\pi(x)x^r \, dx = 0, \quad r = 0, 1, 2, \ldots, n - 1$$

and

$$\int_{-1}^{1} w(x)y \, dx = A_1 y_1 + A_2 y_2 + \cdots + A_n y_n$$

When $n = 2$ and $w(x) = 1$, we find that the polynomial $(x^2 - \frac{1}{3})$ meets our requirements, so that $x_1 = -1/\sqrt{3}$, $x_2 = 1/\sqrt{3}$. We have to find A_1 and A_2 such that

$$\int_{-1}^{1} y \, dx = A_1 y(-1/\sqrt{3}) + A_2 y(1/\sqrt{3})$$

This is exact for $y = 1$ if $A_1 + A_2 = 2$, and for $y = x$ if $A_1 = A_2$. Thus

$$\int_{-1}^{1} y \, dx = y(-1/\sqrt{3}) + y(1/\sqrt{3})$$

The error will be

$$\int_{-1}^{1} (x^2 - \tfrac{1}{3})[x_1, x_2, x] \, dx$$

$$= -\tfrac{1}{3} \int_{-1}^{1} (x^3 - x)[x_1, x_2, x, x] \, dx$$

$$= \tfrac{1}{6} \int_{-1}^{1} (x^2 - 1)^2[x_1, x_2, x, x, x] \, dx$$

$$= \tfrac{8}{45}[x_1, x_2, \xi, \xi, \xi] = y^{(4)}(\xi_1)/135, \quad \text{where } -1 < \xi_1 < 1$$

For $n = 3$, we find $\pi(x) = x^3 - \frac{3}{5}x$, leading to

$$\int_{-1}^{1} y \, dx = \tfrac{1}{9}(5y_1 + 8y_2 + 5y_3)$$

where $x_1 = -\sqrt{\tfrac{3}{5}}$, $x_2 = 0$, $x_3 = \sqrt{\tfrac{3}{5}}$. The error term here is $y^{(6)}(\xi)/15{,}750$.
  For $n = 4$, we find $\pi(x) = x^4 - 6x^2/7 + \tfrac{3}{35}$, leading to

$$\int_{-1}^{1} y \, dx = A_1 y_1 + A_2 y_2 + A_3 y_3 + A_4 y_4$$

where $\quad x_1 = -0{\cdot}8611, \qquad x_2 = -0{\cdot}3400$

$$x_3 = \quad 0{\cdot}3400, \qquad x_4 = \quad 0{\cdot}8611$$

$$A_1 = A_4 = 0{\cdot}3479, \quad A_2 = A_3 = 0{\cdot}6521$$

We can show as above that the error term is $y^{(8)}(\xi)/3{,}472{,}875$ by using the fact that $(x^4 - 6x^2/7 + \tfrac{3}{35})$ is the fourth derivative of $(x^2 - 1)^4/1680$.

  The polynomials $x^3 - 3x/5$ (for $n = 3$) and $x^4 - 6x^2/7 + \tfrac{3}{35}$ (for $n = 4$) are in fact constant multiples of the Legendre polynomials of degrees 3 and 4, of which more later.

  As with Chebyshev integration formulae, the accuracy is gained at the expense of the time taken to evaluate y at the required values of x. However this is no disadvantage when a digital computer is being employed, and the Gaussian integration formulae are today extensively used in programming.

EXAMPLES

(1) Evaluate $\int_{-1}^{1} \sqrt{(1 - x^2)} \, dx$ using (a) the four-point Chebyshev formula (b) the four-point Gauss formula.

(2) Evaluate $\int_{-1}^{1} x^6 \, dx$ by the three-point and by the four-point Chebyshev formula and compare the error with error terms given above.

(3) Calculate $\int_{-1}^{1} x^8 \, dx$ and $\int_{-1}^{1} (1 + x^8)^{\frac{1}{2}} \, dx$ by the four-point Gauss formula.

(4) Find values of x_1, x_2 so that the approximate integration formula

$$\int_{0}^{\pi} y(x) \sin x \, dx = y(x_1) + y(x_2)$$

is exact when y is a cubic polynomial. Use the formula to evaluate the integral when $y(x) = x^4$, and compare the result with the value found by integration by parts.

5

Differential equations

The analytic solution of the equation $dy/dx = 2y$, given $y(0) = 1$, is of course $y = e^{2x}$, but the numerical solution will take the form of a table of values such as the following.

$$
\begin{aligned}
y(0) \quad &= 1, & y(0.3) &= 1.8221 \\
y(0.1) &= 1.2214, & y(0.4) &= 2.2255 \\
y(0.2) &= 1.4918, & y(0.5) &= 2.7183
\end{aligned}
$$

We have to construct this table solely from the equation $dy/dx = 2y$, with the given condition $y(0) = 1$. This involves a choice of the size of the step-length h—in this table it is 0.1—and also a choice of the number of decimal places to be used in the calculation. We set out to evaluate y for a number of values of x, not to find a neat expression for y in terms of x. At each stage, errors due to round-off and to the truncation of series will inevitably creep into the calculation, and if care is not taken, the propagation of the errors will prove disastrous. The key point in any solution must be the control of the growth of error. This is over and beyond repeated checks to ensure that no blunder is overlooked.

First-order equations

The equation $dy/dx = f(x, y)$ gives us the value of the gradient dy/dx at each point (x, y) in the plane. If we give dy/dx a constant value c, the locus of points at which the gradient takes this value is the curve $f(x, y) = c$. Such a curve is called an isoclinal. For the equation $dy/dx = x^2 + y^2$, the isoclinals will be the circles $x^2 + y^2 = c$. For the equation $dy/dx = x - y^2$, they will be the parabolas $y^2 = x - c$. To solve the equation we need to know the value of y for one value of x, i.e., we need to know one point on the graph of the solution. This graph is known as the integral curve corresponding to the given initial condition. We can sketch the integral curve through a given point by means of the isoclinals.

For the equation $dy/dx = x^2 + y^2$, given $y(1) = 0$, the integral curve through the point $(1, 0)$ will cross the isoclinal $x^2 + y^2 = 1$ at $45°$, and then cross the isoclinal $x^2 + y^2 = 2$ at an angle $\tan^{-1} 2$ to the x-axis. By sketching a number of integral curves by hand we can see whether they tend to come closer together or to grow wider apart as x increases. If they tend to come closer together, the tendency for the error in the solution to be magnified is less than if the integral curves grow wider apart. In the latter case, a small error near $x = 0$ will move the solution from one integral curve to a neighbouring integral curve, and the two curves move apart as x increases.

Consider first a very crude step-by-step method for the solution of the equation $dy/dx = 2y$, given $y(0) = 1$, for $x = 0(0\cdot1)\ 0\cdot5$.

We have approximately

$$y_1 = y_0 + hy_0' = 1 + (0\cdot1)(2) = 1\cdot2$$

A better approximation is given by $y_1 = y_0 + h(y_0' + y_1')/2$. If $y_1 = 1\cdot2$, $y_1' = 2\cdot4$ so that

$$y_1 = 1 + (0\cdot1)(2 + 2\cdot4)/2 = 1\cdot22$$

If $y_1 = 1\cdot22$, $y_1' = 2\cdot44$ and this gives us $y_1 = 1\cdot222$. We will work to three decimal places, and the third decimal place in $1\cdot222$ will not be altered if we substitute again. Next we have

$$y_2 = y_1 + hy_1' = 1\cdot222 + (0\cdot1)(2\cdot444) = 1\cdot466$$

We improve this by putting

$$y_2 = y_1 + h(y_1' + y_2')/2$$
$$= 1\cdot222 + (0\cdot1)(2\cdot444 + 2\cdot932)/2 = 1\cdot491$$

By repeating this step we find the value $1\cdot493$, and then $1\cdot494$. Proceeding in this way we obtain the table of values below:

x	y
0·1	1·222
0·2	1·494
0·3	1·826
0·4	2·232
0·5	2·728

At each step we have used the approximation

$$y_{r+1} = y_r + h(y_r' + y_{r+1}')/2$$

Since $y' = 2y$, this gives $y_{r+1} = y_r + (0\cdot1)(y_r + y_{r+1})$, i.e., $y_{r+1} = 11y_r/9$, so that as $y_0 = 1$, $y_n = (\frac{11}{9})^n$. It follows that the truncation error in the final value 2·728 is the difference between $(\frac{11}{9})^5$ and the exact value which we happen to know to be e. In a practical case, we do not know the exact solution, and we have to look elsewhere for a guide to the size of the truncation error.

Since
$$y'_{r+1} = y'_r + hy''_r + h^2 y'''_r/2! + \cdots$$

our approximation for y_{r+1} equals

$$y_r + hy'_r + \tfrac{1}{2}h^2 y''_r + \tfrac{1}{4}h^3 y'''_r + \cdots$$

so that by comparison with the Taylor series for y_{r+1} the truncation error is approximately $-h^3 y'''_r/12$. From $y' = 2y$ we have $y''' = 2y'' = 4y' = 8y$. Hence the truncation error in finding y_{r+1} is approximately $-(0·001)(8y_r/12)$. This is on the assumption that the value of y_r is correct. In fact the error in y_r itself leads to a still greater error in y_{r+1}. We can improve the accuracy by using a smaller step-length h, or by including more terms in the expression for y_{r+1}.

Methods for the solution of first-order differential equations can be divided into two classes

(a) the Runge–Kutta methods and the Taylor series method, which require no special starting procedure. They are one-step methods, in that the calculation of y_{n+1} is based on the values of y and its derivatives at $x = x_n$ or at points between x_n and x_{n+1}.

(b) the predictor-corrector methods and central difference methods which require a set of consecutive values of y from which to begin operations. These starting values can be found by Picard's method or by a one-step method.

Runge–Kutta methods for the equation $dy/dx = f(x, y)$

(a) Put
$$y_{n+1} = y_n + \tfrac{1}{2}k_0 + \tfrac{1}{2}k_1$$

where
$$k_0 = hf(x_n, y_n)$$
$$k_1 = hf(x_n + h, y_n + k_0)$$

This resembles the method used above for the equation $dy/dx = 2y$, except that the value of y_{n+1} is obtained without iteration.

Take $h = 0·05$, $f(x, y) = 2y$, $y(0) = 1$. We have for the first step

$$k_0 = 2h, \quad k_1 = h(2 + 4h)$$
$$y_1 = y_0 + h + h(1 + 2h) = 1·105$$

For the second step

$$k_0 = h(2·210), \quad k_1 = h(2·210 + 4·420h)$$
$$y_2 = 1·105 + 0·055 + 0·061 = 1·221$$

As above, the truncation error is of the order of h^3.

(b) Put

$$y_{n+1} = y_n + \tfrac{1}{6}(k_0 + 4k_1 + k_2)$$

where

$$k_0 = hf(x_n, y_n)$$
$$k_1 = hf(x_n + \tfrac{1}{2}h, y_n + \tfrac{1}{2}k_0)$$
$$k_2 = hf(x_n + h, y_n + 2k_1 - k_0)$$

If we apply this to the equation $dy/dx = \lambda y$, which has the solution $y = A\,e^{\lambda x}$, we find

$$k_0 = h\lambda y_n$$
$$k_1 = h\lambda(y_n + \tfrac{1}{2}h\lambda y_n) = h\lambda y_n(1 + \tfrac{1}{2}h\lambda)$$
$$k_2 = h\lambda(y_n + 2h\lambda y_n + h^2\lambda^2 y_n - h\lambda y_n)$$
$$= h\lambda y_n(1 + h\lambda + h^2\lambda^2)$$

Then

$$y_{n+1} = y_n + \tfrac{1}{6}(h\lambda y_n)(6 + 3h\lambda + h^2\lambda^2)$$
$$= y_n + h\lambda y_n + \tfrac{1}{2}h^2\lambda^2 y_n + \tfrac{1}{6}h^3\lambda^3 y_n$$

i.e.,

$$y_{n+1} = y_n + hy_n' + \tfrac{1}{2}h^2 y_n'' + \tfrac{1}{6}h^3 y_n'''$$

since $\lambda y_n = y_n'$, $\lambda^2 y_n = y_n''$, $\lambda^3 y_n = y_n'''$.

The next term in the Taylor series for y_{n+1} in powers of h is $h^4 y_n^{(4)}/4!$ $= h^4\lambda^4 y_n/24$, and this is a sound estimate of the truncation error. It is thus of the order of h^4.

(c) Put

$$y_{n+1} = y_n + \tfrac{1}{6}(k_0 + 2k_1 + 2k_2 + k_3)$$

where

$$k_0 = hf(x_n, y_n)$$
$$k_1 = hf(x_n + \tfrac{1}{2}h, y_n + \tfrac{1}{2}k_0)$$
$$k_2 = hf(x_n + \tfrac{1}{2}h, y_n + \tfrac{1}{2}k_1)$$
$$k_3 = hf(x_n + h, y_n + k_2)$$

This is the best known of all the Runge–Kutta formulae and is widely used. If we apply it to the equation $dy/dx = \lambda y$ as above, we find that the truncation error is of the order of h^5. Consequently this particular method is spoken of as of fourth-order accuracy. It is worth mentioning that if $f(x, y)$ is independent of y, then both method (b) and method (c) reduce to Simpson's rule.

(d) Put

$$y_{n+1} = y_n + \tfrac{1}{4}(k_0 + 3k_2)$$

where

$$k_0 = hf(x_n, y_n)$$
$$k_1 = hf(x_n + \tfrac{1}{3}h, y_n + \tfrac{1}{3}k_0)$$
$$k_2 = hf(x_n + \tfrac{2}{3}h, y_n + \tfrac{2}{3}k_1)$$

This is known as Heun's method, and has an error term of the order of h^4.

Whichever of these methods is used, some form of check must be applied. This can be done by using another of the methods given, or more commonly by repeating the calculation with a new step-length, usually one half of the previous one. The outstanding advantage of the Runge–Kutta methods is that no preliminary work is required, and for this reason they are called 'self-starting'. Also in their favour is the ease with which it is possible to increase or decrease h when this becomes necessary, and the small demands that the methods make on storage space in a computer. On the other hand it takes time to evaluate $f(x, y)$ at so many intermediate points.

EXAMPLES

(1) If $y' = 1 - 2xy$, $y(0) = 0$, show that $y(0\cdot1) = 0\cdot099336$, $y(0\cdot2) = 0\cdot194751$.
(2) If $y' = -2y$, $y(0) = 1$, calculate to two decimal places the values of y for $x = 0(0\cdot1)0\cdot5$.
(3) If $y' = x - y$, $y(0) = 1$, calculate to two decimal places the values of y for $x = 0(0\cdot2)1\cdot0$ and $x = 0(0\cdot1)1\cdot0$.
(4) Verify that in method (c) above the truncation error is of the order of h^5.

We now consider methods for obtaining values of y near $x = x_0$.

Picard's method

If $dy/dx = f(x, y)$ and $y = y_0$ at $x = 0$ we have

$$y = y_0 + \int_0^x f(x, y) \, dx$$

We cannot integrate this at once since the integrand contains y. To overcome this difficulty we form a sequence $y^{(1)}, y^{(2)}, y^{(3)}, \ldots$, putting

$$y^{(1)} = y_0 + \int_0^x f(x, y_0) \, dx$$

$$y^{(2)} = y_0 + \int_0^x f(x, y^{(1)}) \, dx$$

$$\cdots \cdots \cdots \cdots \cdots$$

$$y^{(i+1)} = y_0 + \int_0^x f(x, y^{(i)}) \, dx$$

Provided that h is sufficiently small, this iteration will converge when $f(x, y)$ is continuous. The method is valuable as a starting procedure, i.e., as a means of finding the value of y at two or three points to form a basis for another method. As an illustration consider the equation

$$\mathrm{d}y/\mathrm{d}x = x \sin(\pi y), \quad y = \tfrac{1}{2} \text{ at } x = 0$$

$$y = \tfrac{1}{2} + \int_0^x x \sin(\pi y)\,\mathrm{d}x$$

$$y^{(1)} = \tfrac{1}{2} + \int_0^x x \sin(\tfrac{1}{2}\pi)\,\mathrm{d}x = \tfrac{1}{2} + \tfrac{1}{2}x^2$$

$$y^{(2)} = \tfrac{1}{2} + \int_0^x x \sin \tfrac{1}{2}\pi(1 + x^2)\,\mathrm{d}x$$

$$= \tfrac{1}{2} + \int_0^x x \cos(\tfrac{1}{2}\pi x^2)\,\mathrm{d}x$$

$$= \tfrac{1}{2} + \int_0^x x(1 - \pi^2 x^4/8 + \cdots)\,\mathrm{d}x$$

$$= \tfrac{1}{2} + \tfrac{1}{2}x^2 - \pi^2 x^6/48 + \cdots$$

We find then that $y^{(3)}$ agrees with $y^{(2)}$ up to and including the term in x^6. We conclude that

$$y = \tfrac{1}{2} + \tfrac{1}{2}x^2 - \pi^2 x^6/48 + \cdots$$

The safest plan is to use the relation $y = \tfrac{1}{2} + \tfrac{1}{2}x^2$ with the knowledge that the error is approximately $-x^6/5$. Thus we can find y_1 and y_2 correct to four decimal places with $h = 0{\cdot}1$, but not y_3.

We can carry out the Picard iteration numerically, and this is a very useful alternative in cases when the analytic integration is difficult. We have the choice between a forward difference method and a Lagrangian method.

By integrating the Newton forward difference formula with respect to p we find

$$\int_0^h y\,\mathrm{d}x = h(y_0 + \tfrac{1}{2}\Delta y_0 - \tfrac{1}{12}\Delta^2 y_0 + \tfrac{1}{24}\Delta^3 y_0 - \cdots)$$

We can now replace y by y' and obtain

$$y_1 - y_0 = h(y_0' + \tfrac{1}{2}\Delta y_0' - \tfrac{1}{12}\Delta^2 y_0' + \tfrac{1}{24}\Delta^3 y_0' - \cdots)$$

By integrating from $x = 0$ to $x = 2h$ we obtain

$$y_2 - y_0 = h(2y_0' + 2\Delta y_0' + \tfrac{1}{3}\Delta^2 y_0' + 0\Delta^3 y_0' - \cdots)$$

and similarly

$$y_3 - y_0 = h(3y_0' + \tfrac{9}{2}\Delta y_0' + \tfrac{9}{4}\Delta^2 y_0' + \tfrac{3}{8}\Delta^3 y_0' + \cdots)$$

We begin by estimating the values of y_1, y_2, and y_3. With these estimates, we use $y' = f(x, y)$ to evaluate y_1', y_2', and y_3' and draw up a difference table for y'. We then use the differences in the equations above to obtain improved approximations for y_1, y_2, and y_3. This is repeated until two successive iterations are in agreement.

Consider the equation

$$dy/dx = x^2 + y^2, \quad y(0) = 1, h = 0.02$$

Since $y'_0 = 1$ we take as our first estimates

$$y_1 = 1.02, \qquad y_2 = 1.04, \qquad y_3 = 1.06$$

From $y' = x^2 + y^2$ we obtain

$$y'_1 = 1.0408, \qquad y'_2 = 1.0832, \qquad y'_3 = 1.1272$$

The difference table for y' now gives

$$\Delta y'_0 = 0.0408, \qquad \Delta^2 y'_0 = 0.0016, \qquad \Delta^3 y'_0 = 0$$

The formulae above give

$$y_1 = 1.0204, \qquad y_2 = 1.0416, \qquad y_3 = 1.0637$$

We now repeat the process.

$$y'_1 = 1.0416, \qquad y'_2 = 1.0864, \qquad y'_3 = 1.1351$$
$$\Delta y'_0 = 0.0416, \qquad \Delta^2 y'_0 = 0.0032, \qquad \Delta^3 y'_0 = 0.0007$$
$$y_1 = 1.0204, \qquad y_2 = 1.0417, \qquad y_3 = 1.0639$$

Now a third iteration

$$y'_1 = 1.0416, \qquad y'_2 = 1.0867, \qquad y'_3 = 1.1355$$
$$\Delta y'_0 = 0.0416, \qquad \Delta^2 y'_0 = 0.0034, \qquad \Delta^3 y'_0 = 0.0005$$
$$y_1 = 1.0204, \qquad y_2 = 1.0417, \qquad y_3 = 1.0639$$

The next iteration produces no change in the values.

For use on a computer, we can express the forward difference formulae above in terms of ordinates, or else obtain the following corresponding Lagrangian formulae by integrating the four-point interpolation polynomial based on the points $x = 0, h, 2h, 3h$.

$$y_1 - y_0 = \tfrac{1}{24}h(9y'_0 + 19y'_1 - 5y'_2 + y'_3)$$
$$y_2 - y_0 = \tfrac{1}{3}h(y'_0 + 4y'_1 + y'_2)$$
$$y_3 - y_0 = \tfrac{3}{8}h(y'_0 + 3y'_1 + 3y'_2 + y'_3)$$

With the same example

$$dy/dx = x^2 + y^2, \quad y(0) = 1, h = 0.02$$

we find, with the same initial estimates,

$$y_1 = 1\cdot02, \qquad y_2 = 1\cdot04, \qquad y_3 = 1\cdot06$$
$$(a)\ y_1' = 1\cdot0408, \qquad y_2' = 1\cdot0832, \qquad y_3' = 1\cdot1272$$

$$y_1 = 1\cdot0204, \qquad y_2 = 1\cdot0416, \qquad y_3 = 1\cdot0637$$
$$(b)\ y_1' = 1\cdot0416, \qquad y_2' = 1\cdot0864, \qquad y_3' = 1\cdot1351$$

$$y_1 = 1\cdot0204, \qquad y_2 = 1\cdot0417, \qquad y_3 = 1\cdot0639$$
$$(c)\ y_1' = 1\cdot0416, \qquad y_2' = 1\cdot0867, \qquad y_3' = 1\cdot1355$$
$$y_1 = 1\cdot0204, \qquad y_2 = 1\cdot0417, \qquad y_3 = 1\cdot0639$$

No alteration is made by the next iteration.

Taylor series method

If y_0 is known, then from $dy/dx = f(x, y)$ we can at once find the value of y_0'. By differentiating we obtain

$$\frac{d^2y}{dx^2} = \frac{\partial f}{\partial x} + \frac{\partial f}{\partial y}\frac{dy}{dx}$$

and by substitution in this expression we find the value of y_0''. Proceeding in this way we evaluate successive derivatives of y at $x = 0$. This enables us to write down the Taylor series

$$y = y_0 + xy_0' + x^2y_0''/2 + x^3y_0'''/6 + \cdots$$

If h is small enough, we can now evaluate y_1, y_{-1}, y_2, y_{-2} from the equations

$$y_1 = y_0 + hy_0' + h^2y_0''/2 + h^3y_0'''/6$$
$$y_{-1} = y_0 - hy_0' + h^2y_0''/2 - h^3y_0'''/6$$
$$y_2 = y_0 + 2hy_0' + 2h^2y_0'' + 4h^3y_0'''/3$$
$$y_{-2} = y_0 - 2hy_0' + 2h^2y_0'' - 4h^3y_0'''/3$$

It is imperative that we verify that it is safe to neglect the first term omitted,

i.e.,   the term $\quad h^4y_0^{(4)}/24$,   when $x = h$

and   the term $\quad 2h^4y_0^{(4)}/3$,   when $x = 2h$

If this is not so, then we must include more terms or else use a smaller value of h. When the series is an alternating series, the first neglected term gives an overestimate of the truncation error. In other cases, especially if the series converges slowly, the first neglected term may give an underestimate of the error.

    As we increase the value of x, we are going to require more and more terms of the series. Since the higher derivatives usually become difficult to evaluate, this means that the Taylor series which we have found is useful only near the origin. Consider again the equation

$$\mathrm{d}y/\mathrm{d}x = x^2 + y^2, \quad y(0) = 1$$

We have $y_0' = 1$, and since $y'' = 2x + 2yy'$, $y_0'' = 2$. Also since $y''' = 2 + 2(y')^2 + 2yy''$, $y_0''' = 8$, and since $y^{(4)} = 6y'y'' + 2yy'''$, $y_0^{(4)} = 28$.

Then
$$y = 1 + x + x^2 + 4x^3/3 + 7x^4/6 + \cdots$$

Since x^4 is $0 \cdot 0001$ when $x = 0 \cdot 1$ we can use the first four terms of the series to evaluate y_{-1} and y_1 with $h = 0 \cdot 1$, correct to three decimal places. If we need y_2 to the same accuracy we shall need to find more terms in the series.

EXAMPLES

(1) Apply Picard's method to the equation

$$\mathrm{d}y/\mathrm{d}x = x^2 + y^2, \quad \text{given } y(0) = 1$$

(2) If $\mathrm{d}y/\mathrm{d}x = x^2 - y$, and $y(0) = 1$, obtain the Taylor series for y in powers of x and hence show that, when $h = 0 \cdot 1$, $y_1 = 0 \cdot 905$, $y_2 = 0 \cdot 821$, $y_3 = 0 \cdot 749$ (correct to three decimal places).
(3) Use forward differences to find y_1, y_2, and y_3 by iteration when $y' = y^2 - x^2$, $y(0) = 1$, with $h = 0 \cdot 02$.

We now turn to the problem of continuing the solution of the differential equation farther and farther from the starting point (x_0, y_0). One obvious possibility is to employ Taylor series. If we form the Taylor series in powers of $(x - x_0)$, we can evaluate $y_1, y_1', y_1'', y_1''', \ldots$. We can now form the Taylor series in powers of $(x - x_1)$ and evaluate $y_2, y_2', y_2'', y_2''', \ldots$. By continuing in this way we can move safely farther from x_0. A convenient and compact notation is used in this process. We write $\tau_r^{(n)} = h^n y_r^{(n)}/n!$, $\tau_r^{(n)}$ being known as a reduced derivative.

The series
$$y_1 = y_0 + hy_0' + h^2 y_0''/2! + h^3 y_0'''/3! + \cdots$$

is written in the form
$$y_1 = (y_0 + \tau_0^{(2)} + \tau_0^{(4)} + \cdots) + (\tau_0^{(1)} + \tau_0^{(3)} + \tau_0^{(5)} + \cdots)$$
Also
$$y_{-1} = (y_0 + \tau_0^{(2)} + \tau_0^{(4)} + \cdots) - (\tau_0^{(1)} + \tau_0^{(3)} + \tau_0^{(5)} + \cdots)$$
From
$$y_1' = y_0' + hy_0'' + h^2 y_0'''/2! + h^3 y_0^{(4)}/3! + \cdots$$
we have
$$\tau_1^{(1)} = (\tau_0^{(1)} + 3\tau_0^{(3)} + 5\tau_0^{(5)} + \cdots) + (2\tau_0^{(2)} + 4\tau_0^{(4)} + \cdots)$$
and
$$\tau_{-1}^{(1)} = (\tau_0^{(1)} + 3\tau_0^{(3)} + 5\tau_0^{(5)} + \cdots) - (2\tau_0^{(2)} + 4\tau_0^{(4)} + \cdots)$$

It is very useful to obtain when possible a recurrence relation between successive reduced derivatives with the same suffix. To illustrate this, consider the example

$$y' = 1 - 2xy, \quad y(0) = 0, h = 0{\cdot}2$$

Then $\tau_0^{(1)} = hy_0' = 0{\cdot}2$, and from $y'' = -2xy' - 2y$ we have $y_0'' = 0$, $\tau_0^{(2)} = 0$.
By differentiating the differential equation $(n - 1)$ times we obtain

$$y^{(n)} = -2xy^{(n-1)} - 2(n - 1)y^{(n-2)} \quad (n \geqslant 2)$$
$$y_0^{(n)} = -2(n - 1)y_0^{(n-2)}$$
$$n\tau_0^{(n)} = -2h^2\tau_0^{(n-2)}$$

Since $\tau_0^{(2)} = 0$, it follows that

$$\tau_0^{(4)} = \tau_0^{(6)} = 0$$

Since $\tau_0^{(1)} = 0{\cdot}2$, we find

$$3\tau_0^{(3)} = -0{\cdot}016, \qquad \tau_0^{(3)} = -0{\cdot}005333$$
$$5\tau_0^{(5)} = -0{\cdot}08\tau_0^{(3)}, \qquad \tau_0^{(5)} = 0{\cdot}000085$$
$$7\tau_0^{(7)} = -0{\cdot}08\tau_0^{(5)}, \qquad \tau_0^{(7)} = -0{\cdot}000001$$

Hence we have $y_1 = 0{\cdot}194751$, $\tau_1^{(1)} = 0{\cdot}184420$.
As a check at this point we can use the relation

$$\tau_1^{(1)} = hy_1' = h(1 - 2x_1y_1), \quad \text{i.e.,} \quad 5\tau_1^{(1)} = 1 - 0{\cdot}4y_1$$

In the next stage, we find the reduced derivatives with suffix 1.

$$2\tau_1^{(2)} = h^2y_1^{(2)} = h^2(-2x_1y_1' - 2y_1)$$
$$\tau_1^{(2)} = -h^2(\tau_1^{(1)} + y_1) = -0{\cdot}015167$$

From
$$y^{(n)} = -2xy^{(n-1)} - 2(n - 1)y^{(n-2)},$$
$$n\tau_1^{(n)} = -0{\cdot}08(\tau_1^{(n-1)} + \tau_1^{(n-2)}) \quad (n \geqslant 2)$$

We use this to find

$$\tau_1^{(3)} = -0{\cdot}004513, \qquad \tau_1^{(4)} = 0{\cdot}000394, \qquad \tau_1^{(5)} = 0{\cdot}000066$$
$$\tau_1^{(6)} = -0{\cdot}000006, \qquad \tau_1^{(7)} = -0{\cdot}000001$$

These values give

$$y_1 + \tau_1^{(2)} + \tau_1^{(4)} + \tau_1^{(6)} = 0{\cdot}179972$$
$$\tau_1^{(1)} + \tau_1^{(3)} + \tau_1^{(5)} + \tau_1^{(7)} = 0{\cdot}179972$$

From these we find

$$y_2 = 0{\cdot}359944, \quad \text{and} \quad y_0 = 0 \quad \text{(Check)}$$

Also
$$\tau_1^{(1)} + 3\tau_1^{(3)} + 5\tau_1^{(5)} + 7\tau_1^{(7)} = 0{\cdot}171205$$
$$2\tau_1^{(2)} + 4\tau_1^{(4)} + 6\tau_1^{(6)} = -0{\cdot}028797$$

From these values we find

$$\tau_{12}^{(1)} = 0{\cdot}171205 - 0{\cdot}028797 = 0{\cdot}142408$$
$$\tau_{0}^{(1)} = 0{\cdot}171205 + 0{\cdot}028797 = 0{\cdot}200002 \quad \text{(Check)}$$

(The exact value of $\tau_{0}^{(1)}$ is $0{\cdot}2$.)

In the next stage we find the reduced derivatives with suffix 2.

$$y_2 = 0{\cdot}359944, \qquad x_2 = 0{\cdot}4$$
$$\tau_{2}^{(2)} = -0{\cdot}04(y_2 + 2\tau_{2}^{(1)}) = -0{\cdot}025790(4)$$

The recurrence relation gives us

$$n\tau_{2}^{(n)} = -0{\cdot}08(2\tau_{2}^{(n-1)} + \tau_{2}^{(n-2)}) \quad (n \geqslant 2)$$
$$\tau_{2}^{(3)} = -0{\cdot}002422, \qquad \tau_{2}^{(4)} = 0{\cdot}000613$$
$$\tau_{2}^{(5)} = 0{\cdot}000019, \qquad \tau_{2}^{(6)} = -0{\cdot}000009$$
$$y_2 + \tau_{2}^{(2)} + \tau_{2}^{(4)} + \tau_{2}^{(6)} = 0{\cdot}334758$$
$$\tau_{2}^{(1)} + \tau_{2}^{(3)} + \tau_{2}^{(5)} = 0{\cdot}140007$$

Thus
$$y_3 = 0{\cdot}474765, \quad \text{and} \quad y_1 = 0{\cdot}194751 \quad \text{(Check)}$$
$$\tau_{2}^{(1)} + 3\tau_{2}^{(3)} + 5\tau_{2}^{(5)} = 0{\cdot}135238$$
$$2\tau_{2}^{(2)} + 4\tau_{2}^{(4)} + 6\tau_{2}^{(6)} = -0{\cdot}049182$$
$$\tau_{3}^{(1)} = 0{\cdot}135238 - 0{\cdot}049182 = 0{\cdot}086056$$
$$\tau_{1}^{(1)} = 0{\cdot}135238 + 0{\cdot}049182 = 0{\cdot}184420 \quad \text{(Check)}$$

The strength of the method lies in the check at the end of each stage provided by recovering earlier values of y and $\tau^{(1)}$.

We now consider a group of methods, known as 'predictor-corrector' methods. These use finite difference formulae or Lagrangian formulae in which the truncation error is under close control. From values of y' at $x = x_n, x_{n-1}, x_{n-2}, \ldots$ we find by means of the predictor formula an estimate of the value of y_{n+1}. This is substituted in the differential equation to obtain an estimate of y'_{n+1}. The open predictor formula is now replaced by a closed corrector formula, and from this a revised estimate for y_{n+1} is found. This is used in the differential equation to determine a revised estimate of y'_{n+1}. If necessary, the corrector formula is used a second and perhaps a third time, until two consecutive estimates coincide.

Adams–Bashforth method

Predictor formula

$$y_{r+1} = y_r + h(1 + \tfrac{1}{2}\nabla + \tfrac{5}{12}\nabla^2 + \tfrac{3}{8}\nabla^3 + \tfrac{251}{720}\nabla^4 + \cdots)y'_r$$

Corrector formula

$$y_{r+1} = y_r + h(1 - \tfrac{1}{2}\nabla - \tfrac{1}{12}\nabla^2 - \tfrac{1}{24}\nabla^3 - \tfrac{19}{720}\nabla^4 - \cdots)y'_{r+1}$$

These two relations have already been obtained by integrating the Newton backward difference formula.

The number of terms to be employed in these formulae will depend on the size of the step-length and on the number of decimal places required.

For the equation

$$y' = x^2 + y^2, \qquad y(0) = 1, \qquad h = 0 \cdot 02$$

we have already found the values

x	y	y'
0	1	1
0·02	1·0204	1·0416
0·04	1·0417	1·0867
0·06	1·0639	1·1355

We draw up a difference table for y'. The values shown in brackets are those obtained in the solution.

y'_0	1			
		416		
y'_1	1·0416		35	
		451		2
y'_2	1·0867		37	
		488		(2)
y'_3	1·1355		(39)	
		(527)		(4)
y'_4	(1·1882)		(43)	
		(570)		
y'_5	(1·2452)			

Then $\qquad \nabla y'_3 = 0 \cdot 0488, \qquad \nabla^2 y'_3 = 0 \cdot 0037, \qquad \nabla^3 y'_3 = 0 \cdot 0002$

We substitute these values in the predictor formula

$$y_4 = y_3 + h\left(y'_3 + \tfrac{1}{2}\nabla y'_3 + \tfrac{5}{12}\nabla^2 y'_3 + \tfrac{3}{8}\nabla^3 y'_3\right)$$

and obtain $y_4 = 1 \cdot 0871$. Then from the differential equation

$$y'_4 = x_4^2 + y_4^2 = (0 \cdot 08)^2 + (1 \cdot 0871)^2 = 1 \cdot 1882$$

This value is entered in the difference table, and gives

$$\nabla y'_4 = 0 \cdot 0527, \qquad \nabla^2 y'_4 = 0 \cdot 0039, \qquad \nabla^3 y'_4 = 0 \cdot 0002$$

The corrector formula

$$y_4 = y_3 + h\left(y'_4 - \tfrac{1}{2}\nabla y'_4 - \tfrac{1}{12}\nabla^2 y'_4 - \tfrac{1}{24}\nabla^3 y'_4\right)$$

then gives $y_4 = 1 \cdot 0871$, in agreement with the predicted value for y_4.

We now use the predictor formula

$$y_5 = y_4 + h(y'_4 + \tfrac{1}{2}\nabla y'_4 + \tfrac{5}{12}\nabla^2 y'_4 + \tfrac{3}{8}\nabla^3 y'_4)$$

and obtain $y_5 = 1{\cdot}1114$. From the differential equation

$$y'_5 = x_5^2 + y_5^2 = (0{\cdot}1)^2 + (1{\cdot}1114)^2 = 1{\cdot}2452$$

The corrector formula then gives $y_5 = 1{\cdot}1114$, as predicted.

Notice that the truncation error in the predictor formula is approximately $251\nabla^4 y'_r/720$ and in the corrector formula is approximately $- 19\nabla^4 y'_{r+1}/720$, both being negligible.

It is useful now to draw up a difference table for values of y, to see that the value is changing smoothly.

$$
\begin{array}{llccc}
y_0 & 1{\cdot}0000 & & & \\
& & 204 & & \\
y_1 & 1{\cdot}0204 & & 9 & \\
& & 213 & & 0 \\
y_2 & 1{\cdot}0417 & & 9 & \\
& & 222 & & 1 \\
y_3 & 1{\cdot}0639 & & 10 & \\
& & 232 & & \\
y_4 & 1{\cdot}0871 & & 11 & \\
& & 243 & & \\
y_5 & 1{\cdot}1114 & & &
\end{array}
$$

In a computer program we would use the same predictor and corrector formulae, expressed in terms of values of y', i.e.,

$$y_{r+1} = y_r + h(55y'_r - 59y'_{r-1} + 37y'_{r-2} - 9y'_{r-3})/24$$

and

$$y_{r+1} = y_r + h(9y'_{r+1} + 19y'_r - 5y'_{r-1} + y'_{r-2})/24$$

The first equation uses known values to predict the value of y_{r+1}. This is used in the differential equation to give a value for y'_{r+1}. We then use this in the corrector equation and iterate with the corrector equation until the difference between two consecutive values is less than the required tolerance.

Milne's method

This is the simplest of several methods introduced by Milne. For the predictor formula he takes

$$
\begin{aligned}
y_{r+1} &= y_{r-3} + 4h(y'_r - \nabla y'_r + \tfrac{2}{3}\nabla^2 y'_r) \\
&= y_{r-3} + 4h(2y'_r - y'_{r-1} + 2y'_{r-2})/3
\end{aligned}
$$

with a truncation error equal to $14h^5 y^{(5)}(\xi_1)/45$.

The corrector formula used is

$$y_{r+1} = y_{r-1} + h(2y'_{r+1} - 2\nabla y'_{r+1} + \tfrac{1}{3}\nabla^2 y'_{r+1})$$
$$= y_{r-1} + \tfrac{1}{3}h(y'_{r+1} + 4y'_r + y'_{r-1})$$

with a truncation error equal to $-h^5 y^{(5)}(\xi_2)/90$. Notice that the truncation errors are of opposite sign and are approximately in the ratio of 28 to 1. It must be emphasized that the truncation error in the corrector formula should be negligible, and that the main error in the first corrected value of y_{r+1} arises from the fact that we are using an estimate of y'_{r+1} on the right-hand side.

We now apply this method to the equation

$$y' = 1 - 2xy, \quad y(0) = 0$$

assuming that the following values have already been found:

	x	y	y'
x_{-1}	-0.2	-0.19475	0.92210
x_0	0	0	1
x_1	0.2	0.19475	0.92210
x_2	0.4	0.35994	0.71204

The predictor equation gives $y_3 = 0.47244$, and this value substituted in $y'_3 = 1 - (1.2)y_3$ gives $y'_3 = 0.43307$. Four successive applications of the corrector equation give the values

(a) $y_3 = 0.47497,$ $y'_3 = 0.43004$

(b) $y_3 = 0.47477,$ $y'_3 = 0.43028$

(c) $y_3 = 0.47479,$ $y'_3 = 0.43025$

(d) $y_3 = 0.47478,$ $y'_3 = 0.43026$

The error of 19 units in the first corrected value of y_3 arises from the error in the predicted value of y'_3 (281 units) multiplied by $\tfrac{1}{3}h$.

EXAMPLES

(1) Use Milne's method in the example above to find y_4 and y_5.

(2) Solve the equation $y' = 1 - 2xy$, $y(0) = 0$, for $x = 0(0.1)1$, by Milne's method. Work to five decimal places. Sketch the isoclinals and draw the integral curve.

Convergence of the iteration

It is essential that successive applications of the corrector formula produce rapid convergence. For the equation $y' = f(x, y)$, a change δy in an estimate of y_{n+1} produces a change $(\partial f/\partial y)\delta y$ in y'_{n+1}. In Milne's method with the corrector formula

$$y_{n+1} = y_{n-1} + \tfrac{1}{3}h(y'_{n+1} + 4y'_n + y'_{n-1})$$

this change in y'_{n+1} produces a change $\frac{1}{3}h(\partial f/\partial y)\delta y$ in y_{n+1}. Thus the ratio of consecutive changes in y_{n+1} is $\frac{1}{3}h(\partial f/\partial y)$, and for rapid convergence this must be much smaller than 1 in magnitude. In the example above, $y' = 1 - 2xy$, $\partial f/\partial y = -2x, h = 0\cdot2$ so that when $x = 0\cdot6$ we have $\frac{1}{3}h(\partial f/\partial y) = -0\cdot08$, and the convergence will be rapid. In the Adams–Bashforth method using the corrector formula

$$y_{n+1} = y_n + \tfrac{1}{12}h(5y'_{n+1} + 8y'_n - y'_{n-1})$$

we shall need $(5h/12)\,|\partial f/\partial y| < 1$ for convergence.

Stability

When the computed solution behaves like the true solution of the differential equation, we say that the solution is stable. Stability will depend on three factors, the method, the differential equation, and the step-length. If there is stability for small h but not for large h, we say that there is partial instability. When we can approximate to $f(x, y)$ by a linear function $ax + by + c$, then a simple transformation will bring the equation $y' = f(x, y)$ to the form $y' = \lambda y$. If we apply Milne's corrector to this equation we obtain the difference equation

$$(1 - \tfrac{1}{3}\lambda h)y_{n+1} - \tfrac{4}{3}\lambda h y_n - (1 + \tfrac{1}{3}\lambda h)y_{n-1} = 0$$

It is a little disturbing to find that our first-order differential equation has been exchanged for a second-order difference equation. The difference equation will have two solutions and they cannot both be right. We solve the difference equation by putting $y_n = r^n$, and we take h sufficiently small to make $\lambda^2 h^2$ negligible.

From

$$(1 - \tfrac{1}{3}\lambda h)r^2 - \tfrac{4}{3}\lambda h r - (1 + \tfrac{1}{3}\lambda h) = 0$$

we have approximately

$$r = (1 + \lambda h), \quad \text{and} \quad r = (-1 + \tfrac{1}{3}\lambda h)$$

Then

$$y_n = A(1 + \lambda h)^n + B(-1 + \tfrac{1}{3}\lambda h)^n$$

If $\lambda^2 h^2$ is negligible, this does not differ from

$$y_n = A \exp(n\lambda h) + B(-1)^n \exp(-n\lambda h/3)$$

If $\lambda = 2$, so that the differential equation is $y' = 2y$, the solution we need is a multiple of $\exp(2x)$, i.e.,

$$y_n = \exp(2x_n) = \exp(2nh) \times \text{constant}$$

This agrees with the first term in the solution of the difference equation. The second term is a multiple of $\exp(-2x_n/3)$, and this will tend to decrease as x_n

increases. We say that this second term is a parasitic solution, which in this case ($\lambda = 2$) does not increase in size. This will be true for all positive values of λ.

If λ is negative the parasitic solution $\exp(-n\lambda h/3)$ has now a positive index and it will tend to grow. However by using a small step-length and by keeping guard figures in our working, we can by reducing the round-off error keep the solution under control. This difficulty is an example of 'partial instability', and it arises when $\partial f/\partial y$ is negative.

Consider as an illustration the equation

$$y' = -5y, \quad \text{with } y(0) = 1$$

We take $h = 0\cdot 1$ and we assume that the following values, which agree with the solution $y = e^{-5x}$, have already been obtained

$$y_1 = 0\cdot 607, \qquad y_2 = 0\cdot 368, \qquad y_3 = 0\cdot 223$$
$$y_1' = -3\cdot 033, \qquad y_2' = -1\cdot 839, \qquad y_3' = -1\cdot 116$$

The predictor formula gives

$$y_4 = y_0 + (0\cdot 4)(2y_3' - y_2' + 2y_1')/3 = 0\cdot 138(8)$$

Hence $y_4' = -0\cdot 694$

The corrector formula gives

$$y_4 = y_2 + (0\cdot 1)(y_4' + 4y_3' + y_2')/3$$

Now $\qquad\qquad 0\cdot 1(y_4' + 4y_3' + y_2') = -0\cdot 699(7)$

Hence $\qquad\qquad y_4 = 0\cdot 134(8), \qquad y_4' = -0\cdot 674$

The next application of the corrector formula gives

$$y_4 = 0\cdot 135(4), \qquad y_4' = -0\cdot 677$$

and then finally

$$y_4 = 0\cdot 135(3), \qquad y_4' = -0\cdot 676(5)$$

The corrector formula leads to a difference equation with roots $r = 0\cdot 6$ and $r = -1\cdot 2$ approximately, and it can be seen that y_4 is close to $0\cdot 6y_3$.

To avoid the partial instability in Milne's method, Hamming replaced the corrector equation by

$$8y_{n+1} = 9y_n - y_{n-2} + 3h(y_{n+1}' + 2y_n' - y_{n-1}')$$

This new corrector equation has a larger truncation error, $\{-h^5 y^5(\xi)/40\}$, but it gives a stable method for any step-length less than $0\cdot 69/|\partial f/\partial y|$.

In the Adams–Bashforth method the corrector equation is

$$y_{n+1} = y_n + h(y_{n+1}' - \tfrac{1}{2}\nabla y_{n+1}' - \tfrac{1}{12}\nabla^2 y_{n+1}')$$
$$= y_n + h(5y_{n+1}' + 8y_n' - y_{n-1}')/12$$

so that we are faced again with a second-order difference equation with two solutions. If we apply this to our test equation $\mathrm{d}y/\mathrm{d}x = \lambda y$ we find

$$(1 - \tfrac{5}{12}\lambda h)y_{n+1} - (1 + \tfrac{2}{3}\lambda h)y_n + \tfrac{1}{12}\lambda h y_{n-1} = 0$$

We first solve the quadratic

$$(1 - \tfrac{5}{12}\lambda h)r^2 - (1 + \tfrac{2}{3}\lambda h)r + \tfrac{1}{12}\lambda h = 0$$

If $\lambda^2 h^2$ can be neglected, we find $r = 1 + \lambda h$, or $r = \lambda h/12$. Then the general solution of the difference equation is approximately

$$y_n = A(1 + \lambda h)^n + B(\lambda h/12)^n$$

The first term $(1 + \lambda h)^n$ corresponds to the required exponential solution $e^{\lambda x}$, while if λh is small compared with 1 the second term will not give trouble. If we substitute $y_n = r^n$ in the corrector formula we find

$$r^2 - r = (\lambda h/12)(5r^2 + 8r - 1)$$

If we sketch the graphs of $(r^2 - r)$ and $(\lambda h/12)(5r^2 + 8r - 1)$ against r for positive and negative values of λh, we find that the unwanted root has a larger modulus than the wanted root only if $\lambda h < -\tfrac{3}{2}$.

In some cases the parasitic solution cannot be brought under control and then the method is said to show 'strong instability'. This happens when one of the roots of the difference equation has a modulus which is greater than one, even when the step-length is taken very small. The presence of parasitic solutions is the reason why we do not attempt to use a corrector formula of extreme accuracy. A very small truncation error would be too expensive if it were bought at the cost of troublesome spurious roots.

Central difference methods

A longer step-length can be used with these methods than in the methods based on backward differences, but the judgment of the operator plays a larger part in the solution. Having evaluated y_n, we calculate y_{n+1} using estimates of differences that have not yet been found. We then re-evaluate y_{n-1} and correct previous estimates where necessary. This completes one cycle. Methods such as this requiring the exercise of judgment are best suited to work on a desk-machine.

In the following typical method, the formula which fills the role of predictor is based on the integration formula

$$\int_{-h}^{h} y \, \mathrm{d}x = h(2y_0 + \delta^2 y_0/3 - \delta^4 y_0/90 + \cdots)$$

This gives

$$y_{n+1} - y_{n-1} = h(2y_n' + \delta^2 y_n'/3 - \delta^4 y_n'/90 + \cdots)$$

The second formula required will now be established.

From $\qquad hD = 2 \sinh^{-1} \tfrac{1}{2}\delta = \delta(1 - \delta^2/24 + 3\delta^4/640 - \cdots)$

we find $\qquad \delta = (1 - \delta^2/24 + 3\delta^4/640 - \cdots)^{-1}hD$

$$= (1 + \delta^2/24 - 17\delta^4/5760 + \cdots)hD$$

We multiply the right-hand side by $\mu(1 + \tfrac{1}{4}\delta^2)^{-\frac{1}{2}}$, which equals unity, and obtain

$$\delta y_n = \mu(1 - \delta^2/12 + 11\delta^4/270 - \cdots)hDy_n$$

We now multiply by δ^{-1} to obtain the formula required.

$$y_n = h\mu(\delta^{-1}y_n' - \delta y_n'/12 + 11\delta^3 y_n'/720 - \cdots)$$

The interpretation of $\delta^{-1}y_n$ will be dealt with in the example that follows.
Consider the equation

$$y' = 0{\cdot}1x - 0{\cdot}02xy, \quad \text{given } y(0) = 0$$

We take $h = 1$ and assume that the following starting values have already been found:

$$y_1 = y_{-1} = 0{\cdot}050, \quad y_1' = 0{\cdot}099, \quad y_{-1}' = -0{\cdot}099$$
$$y_2 = y_{-2} = 0{\cdot}196, \quad y_2' = 0{\cdot}192, \quad y_{-2}' = -0{\cdot}192$$

We form the difference table for y', with a column on the left for the values of $\delta^{-1}y_{n+\frac{1}{2}}'$.

δ^{-1}	y'	δ	δ^2	δ^3
	$-0{\cdot}192$			
107		93		
	$-0{\cdot}099$		6	
8		99		-6
	0		0	
8		99		-6
	$0{\cdot}099$		-6	
107		93		(-5)
	$0{\cdot}192$		(-11)	
299		(82)		(-4)
	$(0{\cdot}274)$		(-15)	
(573)		(67)		(-4)
	$(0{\cdot}341)$		(-19)	
(914)		(48)		
	$(0{\cdot}389)$			

From $y_0 = \mu\delta^{-1}y_0' - \mu\delta y_0'/12 + 11\mu\delta^3 y_0'/720 - \cdots$ we find $\mu\delta^{-1}y_0' = 0{\cdot}008$.

Since $y_0' = 0$,

$$\delta^{-1}y_{\frac{1}{2}}' = \delta^{-1}y_{-\frac{1}{2}}' = 0\cdot008$$
$$\delta^{-1}y_{\frac{3}{2}}' = \delta^{-1}y_{\frac{1}{2}}' + y_1' = 0\cdot107$$
$$\delta^{-1}y_{\frac{5}{2}}' = \delta^{-1}y_{\frac{3}{2}}' + y_2' = 0\cdot299$$

(a) In the first cycle of operations we find y_3 and we check y_1. We then estimate the value of $\delta^2 y_2$. (In a computer program we could use the value of $\delta^2 y_1$.) If we estimate $\delta^2 y_2'$ to be -12 units, we have

$$y_3 = y_1 + h(2y_2' + \delta^2 y_2'/3) = 0\cdot430$$

We substitute this in the differential equation and find $y_3' = 0\cdot274$. We add this entry to the difference table and find $\delta^2 y_2' = -11$ units. This leaves y_3 and y_3' unchanged, and we find $\delta^{-1}y_{3\frac{1}{2}}' = 573$ units. Also

$$y_1 = \mu\delta^{-1}y_1' - \mu\delta y_1'/12 = 0\cdot0495$$

(b) In the second cycle we find y_4 and check y_2. If we estimate $\delta^2 y_3'$ to be -15 units we have

$$y_4 = y_2 + h(2y_3' + \delta^2 y_3'/3) = 0\cdot739, \qquad y_4' = 0\cdot341$$

The difference table then confirms $\delta^2 y_3' = -15$ units. Also

$$y_2 = \mu\delta^{-1}y_2' - \mu\delta y_2'/12 = 0\cdot196$$

(c) In the third cycle we find y_5 and check y_3. If we estimate $\delta^2 y_4'$ to be -18 units we have

$$y_5 = y_3 + h(2y_4' + \delta^2 y_4'/3) = 1\cdot106, \qquad y_5' = 0\cdot389$$

In the difference table we find $\delta^2 y_4' = -19$ units. Also

$$y_3 = \mu\delta^{-1}y_3' - \mu\delta y_3'/12 = 0\cdot430$$

EXAMPLES

(1) Show that

$$h\delta^{-1}y_{\frac{1}{2}}' = \mu y_{\frac{1}{2}} - \tfrac{1}{6}\mu\delta^2 y_{\frac{1}{2}} + \tfrac{1}{30}\mu\delta^4 y_{\frac{1}{2}} - \cdots$$

and use this to check the result $\delta^{-1}y_{\frac{1}{2}}' = 0\cdot008$ in the previous example.

(2) Find the values of $y(0\cdot3)$ and $y(0\cdot4)$, if $y' = -10y$, $h = 0\cdot1$, $y_0 = 1$, $y_1 = 0\cdot368$, and $y_2 = 0\cdot135$.

(3) Use the Taylor series method to solve the equation

$$y' = e^x + xy, \qquad y(0) = 0$$

for $x = 0(0\cdot1)0\cdot3$, giving the values to four decimal places.

(4) Use Picard's method to find $y(2\cdot1)$, $y(2\cdot2)$, $y(2\cdot3)$ given that $xy' = x - y$, $y(2) = 2$. Continue the solution by any method to $x = 2\cdot4$ and $2\cdot5$.

(5) Use the Adams–Bashforth method to solve the equation

$$y' = 1 - 2xy, \quad y(0) = 0$$

with $h = 0.1$, given the values $y_1 = 0.09934$, $y_2 = 0.19475$. Compare the value obtained for $y(0.6)$ with that obtained earlier by the Taylor series method.

(6) Solve the equation $y' = y - x$, (a) with $y(0) = 1$, $h = 0.2$, (b) with $y(0) = 1.1$, $h = 0.2$. Compare your results with those found analytically.

(7) Use Milne's method to find y for $x = 0.3(0.1)1.0$ when

$$y' = (x - 1)^2 - y^2, \quad h = 0.1$$

given the values $y_{-1} = -0.10995$, $y_0 = 0$, $y_1 = 0.09005$, $y_2 = 0.16073$.

(8) If $y' = x + y^2$ and $y(0) = 0$, use a Runge–Kutta method to calculate $y(0.6)$ (a) with $h = 0.2$, (b) with $h = 0.3$.

(9) If $y' = x - y^2$, $y(0) = 0$ obtain starting values with $h = 0.1$ and continue the solution to $x = 0.5$, working with four decimal places.

(10) Use a predictor-corrector method to find $y(0.3)$ and $y(0.4)$ when $y' - y = x - x^2$, $h = 0.1$, given the values $y_0 = 1.200$, $y_1 = 1.331$, $y_2 = 1.484$.

(11) Obtain starting values for the equation $y' = x^2 - y + 1$ given $y(0) = 0$, taking $h = 0.1$. Continue the solution by any method as far as $x = 0.5$, working to four decimal places.

Second-order equations

With first-order equations we needed one initial value of y, i.e., a single condition. With second-order equations we shall need two given conditions. For instance, we may be given the value of y for two values of x, or we may be given the values of y and y' at the starting point.

Consider the equation

$$y'' + f(x)y' + g(x)y = k(x)$$

Suppose that we know the values of y_0 and y_0'. One method of attack is to introduce another variable z defined by $y' = z$.

Then
$$z' = y''(x) = k(x) - f(x)z - g(x)y$$

We now have two simultaneous first-order differential equations to be solved for y and z. Clearly an equation of the third order can in this way be reduced to a system of three simultaneous first-order equations, and the same method is applicable to a linear equation of any order.

We will adapt a Runge–Kutta method, with $h = 0.1$, to solve the equation $y'' + 3xy' - 6y = 0$ given that $y(0) = 1$, $y'(0) = 0.1$. If we put $y' = z$, then $z' = 6y - 3xz$, $z(0) = 0.1$.

(*a*)
$$y_{n+1} = y_n + \tfrac{1}{6}(k_0 + 2k_1 + 2k_2 + k_3)$$

where
$$k_0 = hz_n, \qquad k_1 = h(z_n + \tfrac{1}{2}K_0)$$
$$k_2 = h(z_n + \tfrac{1}{2}K_1), \qquad k_3 = h(z_n + K_2)$$

(*b*)
$$z_{n+1} = z_n + \tfrac{1}{6}(K_0 + 2K_1 + 2K_2 + K_3)$$

where
$$K_0 = h(6y_n - 3x_n z_n)$$
$$K_1 = h\{6(y_n + \tfrac{1}{2}k_0) - 3(x_n + \tfrac{1}{2}h)(z_n + \tfrac{1}{2}K_0)\}$$
$$K_2 = h\{6(y_n + \tfrac{1}{2}k_1) - 3(x_n + \tfrac{1}{2}h)(z_n + \tfrac{1}{2}K_1)\}$$
$$K_3 = h\{6(y_n + k_2) - 3(x_n + h)(z_n + K_2)\}$$

We find

$$k_0 = 0 \!\cdot\! 01, \qquad K_0 = 0 \!\cdot\! 6$$
$$k_1 = 0 \!\cdot\! 04, \qquad K_1 = 0 \!\cdot\! 597$$
$$k_2 = 0 \!\cdot\! 03985, \qquad K_2 = 0 \!\cdot\! 60602$$
$$k_3 = 0 \!\cdot\! 07060, \qquad K_3 = 0 \!\cdot\! 60273$$

and hence we obtain

$$y_1 = 1 \!\cdot\! 0400, \quad \text{and} \quad y_1' = z_1 = 0 \!\cdot\! 7015$$

EXAMPLES

(1) Carry out the calculation for the next step forward, and show that $y_2 = 1 \!\cdot\! 1404$, $y_2' = 1 \!\cdot\! 3059$.

(2) If $y'' + 3xy' - 6y = 0$ and $y(0) = 1$, $y'(0) = 0 \!\cdot\! 1$, obtain the expansion

$$y = 1 + 0 \!\cdot\! 1x + 3x^2 + 0 \!\cdot\! 05x^3 - 0 \!\cdot\! 0075x^5 - 0 \!\cdot\! 0016x^7 - \cdots$$

and hence evaluate y and y' at $x = 0 \!\cdot\! 1$ and $0 \!\cdot\! 2$.

We can continue the solution above by the Runge–Kutta method, or we can adapt the predictor-corrector method for use with two simultaneous equations. We can obtain the following values for y and y' from the series solution and calculate the corresponding values of z'.

x	y	$y' = z$	$y'' = z'$
0	1	0·1	6
0·05	1·0125	0·4004	6·0150
0·10	1·0400	0·7015	6·0298
0·15	1·0827	1·0034	6·0445
0·20	1·1404	1·3059	6·0588

As predictor formulae we can use the relations

$$y_{r+1} = y_{r-3} + (4h/3)(2y_r' - y_{r-1}' + 2y_{r-2}')$$
$$z_{r+1} = z_{r-3} + (4h/3)(2z_r' - z_{r-1}' + 2z_{r-2}')$$

With $r = 4$ and $h = 0 \!\cdot\! 05$ these give $y_5 = 1 \!\cdot\! 2133$, $z_5 = 1 \!\cdot\! 6092$.

With these values, the differential equation gives us $z_5' = 6{\cdot}0729$.

As corrector formulae we will use the relations

$$y_{r+1} = y_{r-1} + \tfrac{1}{3}h(y_{r+1}' + 4y_r' + y_{r-1}')$$
$$z_{r+1} = z_{r-1} + \tfrac{1}{3}h(z_{r+1}' + 4z_r' + z_{r-1}')$$

With $r = 4$, we put $y_5' = z_5 = 1{\cdot}6092$ in the first of this pair, and $z_5' = 6{\cdot}0729$ in the second. We obtain

$$y_5 = 1{\cdot}2133, \quad z_5 = 1{\cdot}6093, \quad z_5' = 6{\cdot}0728$$

A second application of the corrector formula confirms these values. The solution can be continued in this way. Difference tables for y, z, and z' should be drawn up so that an eye can be kept on the smoothness of their values.

A step-by-step solution of this kind, based on given initial values of y and y', is known as a marching problem. When the differential equation is linear, i.e., it is of the form

$$y'' + f(x)y' + g(x)y = k(x)$$

it may be worthwhile to remove the first derivative by means of a transformation. If we put $u = y \exp\left(\tfrac{3}{4}x^2\right)$ in the equation

$$y'' + 3xy' - 6y = 0$$

we obtain

$$4u'' - (9x^2 + 30)u = 0$$

In the general case, we put $u = y \exp\left(\tfrac{1}{2}\int f(x)\,\mathrm{d}x\right)$ to remove the middle term.

We now consider second-order equations of the type

$$y'' = f(x, y), \quad (y_0 \text{ and } y_0' \text{ given})$$

in which the first derivative y' does not appear. Equations of this form occur very frequently in applied mathematics. Our method is based on the relation

$$\delta^2 y_r = h^2(y_r'' + \tfrac{1}{12}\delta^2 y_r'' - \tfrac{1}{240}\delta^4 y_r'' + \cdots)$$

obtained by writing $h^2 D^2 = \delta^2 - \tfrac{1}{12}\delta^4 + \tfrac{1}{90}\delta^6 - \cdots$ in the form

$$\delta^2 y_r = (1 - \tfrac{1}{12}\delta^2 + \tfrac{1}{90}\delta^4 - \cdots)^{-1}h^2 D^2 y_r$$

From $y'' = f(x, y)$, we have

$$\delta^2 y_r = h^2(f_r + \delta^2 f_r/12 - \delta^4 f_r/240 + \cdots)$$

where f_r denotes $f(x_r, y_r)$. Assuming that we have already calculated a sufficient number of starting values, and have drawn up difference tables for y and for $f(x, y)$, we proceed as follows. Let y_r and f_r be known. We estimate $\delta^2 f_r$ and from the relation above we find $\delta^2 y_r$. The term $\delta^4 f_r/240$ will be negligible if a suitable step-length h is chosen. From $\delta^2 y_r$, we can find

y_{r+1}, and then from the differential equation we calculate f_{r+1}. The difference table will now give us a revised estimate of $\delta^2 f_r$. We now repeat the cycle if this is necessary.

To illustrate the method, consider the equation

$$y'' = xy, \quad y(0) = 0, y'(0) = 1$$

with $h = 0.2$. We obtain starting values from the Taylor series

$$y = x + x^4/12 + x^7/504 + \cdots$$
$$y_1 = 0.2001, \quad y_2 = 0.4021, \quad y_3 = 0.6109$$

The difference table for $f(x, y)$ is now drawn up:

$$f_0 = 0$$
$$400$$
$$f_1 = 0.0400 \qquad 809$$
$$1209 \qquad 38$$
$$f_2 = 0.1609 \qquad 847$$
$$2056$$
$$f_3 = 0.3665$$

If we estimate $\delta^2 f_3$ to be 0.0900, we find

$$\delta^2 y_3 = h^2(f_3 + \delta^2 f_3/12 - \delta^4 f_3/240 + \cdots)$$
$$= 0.04(0.3665 + 0.0075) = 0.0150$$

i.e., $y_4 - 2y_3 + y_2 = 0.0150$, and hence $y_4 = 0.8347$.

From $f_4 = x_4 y_4$, we have $f_4 = 0.6678$. We enter this in the difference table, and find $\delta^2 f_3 = 0.0957$. We repeat the cycle and emerge with the values of y_4 and f_4 unchanged. A difference table for the values of y should be drawn up, and the two difference tables provide a running check on the calculation. By multiplying our expression for $\delta^2 y_r$ by δ^2 we can obtain the relation $\delta^4 y_r = h^2(\delta^2 f_r + \delta^4 f_r/12 - \cdots)$, which can be employed as a check when the fourth differences are known. In the example above, $\delta^4 y_2 = 33$ units and $h^2(\delta^2 f_2 + \delta^4 f_2/12) = 34$ units.

From the Taylor series we find y_4 to be 0.83455. The value 0.8347 found above suffers from round-off error, arising from the fact that in the calculation we double the round-off error of nearly $\frac{1}{2} \times 10^{-4}$ in the value 0.6109 for y_3.

For equations of the type $y'' = f(x)y$ we can obtain a powerful method of solution by multiplying the relation

$$h^2 y_r'' = \delta^2 y_r - \delta^4 y_r/12 + \delta^6 y_r/90 - \cdots$$

by $(1 + \delta^2/12)$. This gives us

$$(1 + \delta^2/12)h^2 y_r'' = \delta^2 y_r + \delta^6 y_r/240 + \cdots$$

If the term $\delta^6 y_r/240$ is small enough to be neglected we have the relation

$$\delta^2 y_r = h^2 y_r'' + h^2 \delta^2 y_r''/12$$

Now if $y_r'' = f(x_r)y_r$,

$$\delta^2 y_r'' = f(x_{r+1})y_{r+1} - 2f(x_r)y_r + f(x_{r-1})y_{r-1}$$

Then

$$y_{r+1}\{1 - \tfrac{1}{12}h^2 f(x_{r+1})\} = y_r\{2 + \tfrac{5}{6}h^2 f(x_r)\} - y_{r-1}\{1 - \tfrac{1}{12}h^2 f(x_{r-1})\}$$

This recurrence relation has a very small truncation error (approximately $\delta^6 y_r/240$). If we apply it to the previous example, $y'' = xy$ (in which $f(x) = x$, $h = 0{\cdot}2$), we find

$$y_4\{1 - (0{\cdot}032)/12\} = y_3\{2 + 5(0{\cdot}024)/6\} - y_2\{1 - (0{\cdot}016)/12\}$$

With $y_3 = 0{\cdot}6109$, $y_2 = 0{\cdot}4021$, this gives $y_4 = 0{\cdot}8346$.

EXAMPLES

(1) Continue the solution of the equation $y'' = xy$ with $y(0) = 0$, $y'(0) = 1$, $h = 0{\cdot}2$, and show that $y_5 = 1{\cdot}0853$ by two distinct methods.

(2) Evaluate y to four decimal places at $x = 0(0{\cdot}1)0{\cdot}5$ when

$$y'' = xy' + y, \qquad y(0) = 1, \qquad y'(0) = 0$$

by putting $y' = z$.

(3) Obtain the series for y in powers of x as far as the term in x^7 when

$$y'' = xy + y^2, \qquad y(0) = 0, \; y'(0) = 1$$

and hence find the value of y correct to five decimal places at $x = -0{\cdot}2$ $(0{\cdot}1)0{\cdot}2$. Continue the solution as far as $x = 0{\cdot}5$.

We now have to consider the case in which we are given the value of y at the two end-points of an interval, and are required to solve the second-order differential equation at points in the interval. This is spoken of as a jury problem.

To solve the equation $y'' = xy$ with the boundary conditions $y(0) = a$, $y(1) = b$, we can obtain solutions to the marching problems

(a) $y'' = xy$, with $y(0) = 0$, $y'(0) = 1$,

(b) $y'' = xy$, with $y(0) = 1$, $y'(0) = 0$.

We then form a linear combination of these two solutions to satisfy the given boundary conditions.

This approach is possible only when the differential equation is linear in y and its derivatives. In the case $y'' = y^2$ with boundary conditions $y(0) = 1$, $y(1) = 2$, we could solve first the marching problem

$$y'' = y^2, \quad y(0) = 1, \; y'(0) = 0$$

Having found $y(1)$ under these conditions we repeat the process with a revised value of $y'(0)$. By interpolation we can then obtain a better estimate for $y'(0)$. By repeating this, we can obtain our solution to the accuracy required, except when a small change in the value of y'_0 leads to a large change in the value of $y(1)$.

In most cases the method of deferred correction would be preferable. As the name implies, we begin by finding an approximate solution to the boundary problem and then improve upon it. In fact we use the approximate solution itself to provide its own correction. In the first stage we replace the differential equation by a set of simultaneous algebraic equations. For illustration, we will use the equation

$$y'' = xy$$

with boundary conditions $y(0) = 0$, $y(1) = 1$, and $h = 0\cdot2$. We again employ the relation

$$\delta^2 y_r = h^2(f_r + \delta^2 f_r/12 - \delta^4 f_r/240 + \cdots)$$

The first approximate solution is obtained from the relation

$$\delta^2 y_r = h^2 f_r = h^2 x_r y_r$$

in which the error is given by $h^2(\delta^2 f_r/12 - \delta^4 f_r/240 + \cdots)$. From this relation we have four equations, one for each of the points of subdivision of the interval.

$$y_2 - 2y_1 + y_0 = h^2 x_1 y_1$$
$$y_3 - 2y_2 + y_1 = h^2 x_2 y_2$$
$$y_4 - 2y_3 + y_2 = h^2 x_3 y_3$$
$$y_5 - 2y_4 + y_3 = h^2 x_4 y_4$$

Now $y_0 = 0$, $y_5 = 1$, and $h = 0\cdot2$, so that the equations become

$$y_2 - 2\cdot008y_1 \qquad\quad = 0$$
$$y_3 - 2\cdot016y_2 + y_1 = 0$$
$$y_4 - 2\cdot024y_3 + y_2 = 0$$
$$1 - 2\cdot032y_4 + y_3 = 0$$

By expressing y_2, y_3, and y_4 in terms of y_1 we can show that the solution of this system of simultaneous equations to five significant figures is

$$y_1 = 0\cdot18496, \quad y_2 = 0\cdot37139$$
$$y_3 = 0\cdot56374, \quad y_4 = 0\cdot76956$$

We now use these values to find the values of xy and to draw up the difference table for y''.

$$
\begin{array}{llll}
f_0 & 0 & & \\
 & & 370 & \\
f_1 & 0\cdot0370 & & 746 \\
 & & 1116 & \\
f_2 & 0\cdot1486 & & 780 \\
 & & 1896 & \\
f_3 & 0\cdot3382 & & 878 \\
 & & 2774 & \\
f_4 & 0\cdot6156 & & 1070 \\
 & & 3844 & \\
f_5 & 1\cdot0000 & &
\end{array}
$$

Then to two significant figures the correction terms required are

$$h^2\delta^2 f_1/12 = 0\cdot00025, \qquad h^2\delta^2 f_2/12 = 0\cdot00026$$
$$h^2\delta^2 f_3/12 = 0\cdot00029, \qquad h^2\delta^2 f_4/12 = 0\cdot00036$$

The system of equations is now

$$
\begin{aligned}
y_2 - 2\cdot008y_1 \qquad\quad &= 0\cdot00025 \\
y_3 - 2\cdot016y_2 + y_1 &= 0\cdot00026 \\
y_4 - 2\cdot024y_3 + y_2 &= 0\cdot00029 \\
1 - 2\cdot032y_4 + y_3 &= 0\cdot00036
\end{aligned}
$$

We deduce that the corrections z_1, z_2, z_3, z_4 to the approximate values already found for y_1, y_2, y_3, y_4 will satisfy the system of equations

$$
\begin{aligned}
z_2 - 2\cdot008z_1 \qquad\quad &= 0\cdot00025 \\
z_3 - 2\cdot016z_2 + z_1 &= 0\cdot00026 \\
z_4 - 2\cdot024z_3 + z_2 &= 0\cdot00029 \\
0 - 2\cdot032z_4 + z_3 &= 0\cdot00036
\end{aligned}
$$

Since z_1, z_2, z_3, and z_4 are small, we can simplify this to

$$
\begin{aligned}
z_2 - 2z_1 \qquad\quad &= 0\cdot00025 \\
z_3 - 2z_2 + z_1 &= 0\cdot00026 \\
z_4 - 2z_3 + z_2 &= 0\cdot00029 \\
-2z_4 + z_3 &= 0\cdot00036
\end{aligned}
$$

From the first two equations we have

$$2z_3 - 3z_2 = 0\cdot00077$$

From the last two equations we have

$$-3z_3 + 2z_2 = 0\cdot00094$$

Hence
$$z_2 = -0{\cdot}00084, \qquad z_3 = -0{\cdot}00087$$
$$z_1 = -0{\cdot}00055, \qquad z_4 = -0{\cdot}00061$$

We now revise our approximations for y_1, y_2, y_3, y_4:
$$y_1 = 0{\cdot}18441, \qquad y_2 = 0{\cdot}37054$$
$$y_3 = 0{\cdot}56287, \qquad y_4 = 0{\cdot}76895$$

With these values we draw up a fresh difference table for y'' and find improved values for the correction terms:
$$h^2\delta^2 f_1/12 = 0{\cdot}00025, \qquad h^2\delta^2 f_2/12 = 0{\cdot}00027$$
$$h^2\delta^2 f_3/12 = 0{\cdot}00028, \qquad h^2\delta^2 f_4/12 = 0{\cdot}00036$$

We solve again for z_1, z_2, z_3, and z_4 and find no change in their values (to two significant figures). Hence our values for the correction terms are sufficiently accurate and our result is given by
$$y_1 = 0{\cdot}1844, \qquad y_2 = 0{\cdot}3705$$
$$y_3 = 0{\cdot}5629, \qquad y_4 = 0{\cdot}7690$$

EXAMPLES

(1) Solve by the deferred correction method the problem
$$y'' = x + y, \quad y(0) = 0, \quad y(1) = 1, \text{ with } h = 0{\cdot}2$$

(2) If $y'' = y + x^2 y$, $y(0) = 0$, $y'(0) = 0{\cdot}5$ find the values of y at $x = \pm 0{\cdot}1$ and $\pm 0{\cdot}2$ to five decimal places. Continue the solution as far as $x = 0{\cdot}4$.

(3) Obtain a recurrence relation equivalent to the equation
$$xy'' + y' + xy = 0$$

Hence calculate y at $x = 0{\cdot}1(0{\cdot}1)0{\cdot}5$ given $y(0) = 1$, $y(0{\cdot}1) = 0{\cdot}9975$.

(4) If $y'' = x + xy$, $y(0) = y'(0) = 1$, show that $y(-0{\cdot}1) = 0{\cdot}8997$, $y(0{\cdot}1) = 1{\cdot}1003$. Continue the solution as far as $x = 0{\cdot}5$, using $h = 0{\cdot}1$.

(5) Solve the differential equation $y'' - y = 4x$, given $y(0) = 0$, $y(1) = 1$, using $h = 0{\cdot}25$.

(6) Solve the equation $y'' = 4y$, given $y(0) = 1$, $y(0{\cdot}5) = 2{\cdot}7183$, using $h = 0{\cdot}1$.

(7) If $y'' = xy$ obtain the series for y in powers of x, given that $y(0) = 0$, $y'(0) = 1$. Calculate the values of y for $x = 0(0{\cdot}1)1$ working to five decimal places.

(8) If $y'' + x + x^3 = 0$, $y(0) = 1$, $y'(0) = 0$ evaluate y for $x = 0(0{\cdot}1)1$, working to four decimal places.

(9) Use the method of deferred correction to find y for $x = 0(0{\cdot}25)1$ given $y'' = x + y$, $y(0) = 0$, $y(1) = 1$. Work to three decimal places.

(10) Given $y'' = y + x^2 y$, $y(0) = 0$, $y'(0) = 1$, evaluate y for $x = 0(0{\cdot}1)0{\cdot}5$, working with five decimal places.

(11) Obtain the series for y in powers of x as far as the term in x^9 when $y'' + y^3 = 0$, $y(0) = 0$, $y'(0) = 1$. Evaluate y for $x = 0(0{\cdot}1)1$, working to five decimal places.

(12) If $y'' + y' = y$, find the values of y correct to three decimal places for $x = 1(0{\cdot}5)5$, given $y(1) = 0$, $y(5) = 1$.

We have considered some of the most common methods for the numerical solution of ordinary differential equations, and it must by now be clear that the choice of a method appropriate to a given problem is far from simple. We can distinguish between methods suitable for a hand machine, in which the experience of the operator will be valuable, and methods suitable for an electronic computer, in which each cycle requires a number of iterations. We can balance the advantage of a small step-length, with negligible truncation error, against the advantage of a large step-length, with fewer steps and less round-off error. Above all, we must be on our guard against instability in the solution. We need a wide variety of methods and we need practical experience in their use.

6
Series approximations

In a computer program we may require the value of a function such as $\cos x$ or $\log x$ for various values of x. Hence we need subroutines for the calculation of such values, and for these we need approximations to the functions. These approximations must be capable of being evaluated by simple arithmetic, and they must be of the accuracy required.

We have already considered the question of approximating to a function by means of a polynomial which takes the same value as the function at certain given points. This is known as the method of collocation, and the conditions are satisfied by the appropriate Lagrange interpolation polynomial. When the given points are equally spaced we can form a difference table, and find the polynomial by means of Newton's forward difference formula. Thus the polynomial $4x - 4x^2$ takes the same value as the function $\sin \pi x$ at the points $x = 0, \frac{1}{2}, 1$, but this approximation is not very satisfactory, for the polynomial $4x - 4x^2$ is larger than $\sin \pi x$ in the range $0 < x < 1$ except at the point $x = \frac{1}{2}$.

If the values of the function at the given points are the result of experiment, or if they are rounded values, the advantages of the method of collocation are to some extent lost. If we know values of the function beside the values which we are employing, then clearly we could use them to improve the approximation. For example, we may need to approximate to a function by a quadratic expression in x in a given range, and for this we need at least three values of the function. With three values we can use the method of collocation, but if we have five values it is wise to make use of them all.

Consider the problem of finding a quadratic approximation for the function x^3, using the values of x^3 at $x = 0, 1, 2, 3$, and 4. If the quadratic is $ax^2 + bx + c$, the error in the approximation at any point is

$$\{x^3 - (ax^2 + bx + c)\}$$

and the sum of the squares of the errors at the five chosen points is

$$\sum_{n=0}^{4} \{n^3 - (an^2 + bn + c)\}^2$$

This expression is a function of the three parameters a, b, and c and to find its minimum value we differentiate partially with respect to each parameter in turn and equate to zero.

$$-\sum_{n=0}^{4} 2n^2\{n^3 - (an^2 + bn + c)\} = 0$$

$$-\sum_{n=0}^{4} 2n\{n^3 - (an^2 + bn + c)\} = 0$$

$$-\sum_{n=0}^{4} 2\{n^3 - (an^2 + bn + c)\} = 0$$

These reduce to the equations

$$354a + 100b + 30c = 1300$$
$$100a + 30b + 10c = 354$$
$$30a + 10b + 5c = 100$$

Hence $a = 6$, $b = -8{\cdot}6$, $c = 1{\cdot}2$, so that our quadratic approximation is $6x^2 - 8{\cdot}6x + 1{\cdot}2$. The errors in this approximation at the points $x = 0, 1, 2, 3$, and 4 are $-1{\cdot}2$, $2{\cdot}4$, 0, $-2{\cdot}4$, and $1{\cdot}2$ respectively, giving the sum of the squares of the errors equal to $14{\cdot}4$. This is the smallest possible value for this set of five points, and hence the name 'method of least squares'.

We could do better if we use our knowledge of the value of the function x^3 at all the points of the range, and choose the parameters a, b, and c so that the integral

$$\int_0^4 \{x^3 - (ax^2 + bx + c)\}^2 \, dx$$

is a minimum. For this we require

$$\int_0^4 x^2(x^3 - ax^2 - bx - c) \, dx = 0$$

$$\int_0^4 x(x^3 - ax^2 - bx - c) \, dx = 0$$

and
$$\int_0^4 (x^3 - ax^2 - bx - c) \, dx = 0$$

We solve for a, b, and c and find $a = 6$, $b = -9{\cdot}6$, $c = 3{\cdot}2$. Our approximating quadratic is now $6x^2 - 9{\cdot}6x + 3{\cdot}2$.

Straight line of best fit

If we are looking for a linear function $ax + b$ as an approximation to a function y, when we are given the value of y at the points $x_1, x_2, \ldots, x_n$, we aim at minimizing the sum of the squared errors

$$\sum_{r=1}^{n} (y_r - ax_r - b)^2$$

We equate to zero the partial derivative of this expression with respect to a and also its partial derivative with respect to b. This gives us the 'Normal Equations'

$$\sum_{r=1}^{n} 2x_r(y_r - ax_r - b) = 0$$

$$\sum_{r=1}^{n} 2(y_r - ax_r - b) = 0$$

We write these equations in the form

$$a \sum x_r^2 + b \sum x_r = \sum x_r y_r$$
$$a \sum x_r + nb = \sum y_r$$

and solve for a and b. With these values for a and b we can use the normal equations to show that the sum of the squared errors is equal to

$$\sum_{r=1}^{n} y_r(y_r - ax_r - b)$$

This provides us with a check at the end of the calculation. To show the type of layout employed, we find the line of best fit for the following set of values:

x	1	2	3
y	3·0	5·2	8·0

x_r	x_r^2	y_r	$x_r y_r$	s_r
1	1	3·0	3·0	8·0
2	4	5·2	10·4	21·6
3	9	8·0	24·0	44·0
Total 6	14	16·2	37·4	73·6

The last column gives the row sums and provides a check.

Our normal equations are

$$14a + 6b = 37\cdot4$$
$$6a + 3b = 16\cdot2$$

Hence $a = 2\cdot5$, $b = 0\cdot4$, and the equation of the line of best fit is $y = 2\cdot5x + 0\cdot4$.

As a check the sum of the squared errors is found in two ways

(a) $(0.1)^2 + (0.2)^2 + (0.1)^2 = 0.06$

(b) $\sum y_r(y_r - 2.5x_r - 0.4) = (3.0)(0.1) + (5.2)(-0.2) + 8(0.1) = 0.06$

To find the root-mean-square error we divide 0.06 by the number of points and take the square root. This gives r.m.s. error $= \sqrt{(0.02)} = 0.14$. We conclude that the straight line is not a good fit, but it is the best obtainable.

Parabola of best fit

If we require a quadratic $ax^2 + bx + c$ as an approximation to a function y, we minimize the expression

$$\sum_{r=1}^{n} (y_r - ax_r^2 - bx_r - c)^2$$

By partial differentiation with respect to a, b, and c respectively, we obtain the normal equations

$$\sum_{r=1}^{n} x_r^2(y_r - ax_r^2 - bx_r - c) = 0$$

$$\sum_{r=1}^{n} x_r(y_r - ax_r^2 - bx_r - c) = 0$$

$$\sum_{r=1}^{n} (y_r - ax_r^2 - bx_r - c) = 0$$

We can write these in the form

$$s_4a + s_3b + s_2c = \sum x_r^2 y_r$$
$$s_3a + s_2b + s_1c = \sum x_r y_r$$
$$s_2a + s_1b + nc = \sum y_r$$

where s_i denotes the sum $\sum_{r=1}^{n} x_r^i$, $i = 1, 2, \ldots, n$. We solve these equations for the coefficients a, b, and c, and as a final check we verify that the sum of the squared errors equals $\sum_{r=1}^{n} y_r(y_r - ax_r^2 - bx_r - c)$.

EXAMPLES

(1) Find the straight line of best fit for the pairs of values:

x	-2	-1	0	1	2
y	-4.8	-3.1	-1.0	1.1	2.9

Calculate the r.m.s. error in the result.

(2) Find the parabola of best fit (with equation of the form $y = ax^2 + bx + c$) for the data in the following table:

$$x \quad 0 \qquad 1 \qquad 2 \quad 3 \quad 4$$
$$y \quad -2 \cdot 1 \quad -0 \cdot 4 \quad 2 \cdot 1 \quad 3 \cdot 6 \quad 9 \cdot 9$$

(3) A relation of the form $s = ut + \frac{1}{2}ft^2$ is required between the variables s and t, which take the values:

$$s \quad 10 \quad 30 \quad 50 \quad 80 \quad 110$$
$$t \quad 0 \cdot 9 \quad 1 \cdot 9 \quad 3 \cdot 0 \quad 3 \cdot 9 \quad 5 \cdot 0$$

Show that the method of least squares gives $u = 10 \cdot 7, f = 4 \cdot 63$, and that the sum of the squared errors is then approximately 24.

(4) Find constants a and b such that the curve

$$y = a \cos x + b \sin x$$

fits the following values. Use the result to estimate the maximum value of y.

$$x \quad 0 \quad \pi/6 \quad \pi/4 \quad \pi/3 \quad \pi/2$$
$$y \quad 4 \cdot 0 \quad 5 \cdot 1 \quad 5 \cdot 1 \quad 4 \cdot 8 \quad 3 \cdot 2$$

(5) By taking logarithms and using the method of least squares, find constants a and b so that the curve $y = a\, e^{bx}$ fits the following values:

$$x \quad 0 \qquad 0 \cdot 5 \quad 1 \qquad 1 \cdot 5 \qquad 2 \qquad 2 \cdot 5$$
$$y \quad 10 \cdot 0 \quad 27 \cdot 2 \quad 73 \cdot 9 \quad 201 \cdot 0 \quad 546 \cdot 0 \quad 1484 \cdot 1$$

Legendre polynomials

We consider next the possibility of approximating to a function by a series of polynomials. For use in the range $-1 \leqslant x \leqslant 1$ we have first the Legendre polynomials, which can be generated by the recurrence relation

$$(n + 1)P_{n+1}(x) = (2n + 1)xP_n(x) - nP_{n-1}(x)$$

with $P_0(x) = 1$, $P_1(x) = x$. We find

$$P_2(x) = (3x^2 - 1)/2, \qquad P_3(x) = (5x^3 - 3x)/2$$
$$P_4(x) = (35x^4 - 30x^2 + 3)/8, \quad P_5(x) = (63x^5 - 70x^3 + 15x)/8$$

This set of polynomials is orthogonal with respect to integration over the range $(-1, 1)$, i.e.,

$$\int_{-1}^{1} P_m(x)P_n(x)\, dx = 0, \quad \text{if } m \neq n$$

Also

$$\int_{-1}^{1} P_n^2(x)\, dx = 2/(2n + 1)$$

To find a series $a_0 P_0 + a_1 P_1 + a_2 P_2 + \cdots + a_n P_n$ to use as an approximation to a known function y in the range $(-1, 1)$, we minimize the integral

$$\int_{-1}^{1} \{y - (a_0 P_0 + a_1 P_1 + \cdots + a_n P_n)\}^2 \, \mathrm{d}x$$

by a suitable choice of values for the coefficients $a_0, a_1, \ldots, a_n$. If we differentiate partially with respect to a_r and equate to zero, we obtain

$$\int_{-1}^{1} P_r\{y - (a_0 P_0 + a_1 P_1 + \cdots + a_n P_n)\} \, \mathrm{d}x = 0$$

By the orthogonal property of these polynomials this simplifies to

$$\int_{-1}^{1} P_r\{y - a_r P_r\} \, \mathrm{d}x = 0, \quad r = 0, 1, 2, \ldots, n$$

From this we find

$$a_r = \tfrac{1}{2}(2r + 1) \int_{-1}^{1} y P_r(x) \, \mathrm{d}x$$

It is the orthogonality property of the Legendre polynomials that enables us to obtain such a simple result. Notice that if we wish to improve the accuracy of the approximation by using more terms in the series, we do not have to recalculate the coefficients we have already found.

To illustrate this, we take the function $\cos\left(\tfrac{1}{2}\pi x\right)$, and obtain its expansion in terms of $P_0(x)$, $P_1(x)$, and $P_2(x)$.
We have

$$a_0 = \tfrac{1}{2} \int_{-1}^{1} \cos\left(\tfrac{1}{2}\pi x\right).1 \, \mathrm{d}x = 2/\pi$$

$$a_1 = \tfrac{3}{2} \int_{-1}^{1} \cos\left(\tfrac{1}{2}\pi x\right).x \, \mathrm{d}x = 0$$

$$a_2 = \tfrac{5}{2} \int_{-1}^{1} \cos\left(\tfrac{1}{2}\pi x\right).\tfrac{1}{2}(3x^2 - 1) \, \mathrm{d}x = (10\pi^2 - 120)/\pi^3$$

Our approximation is therefore

$$\cos\left(\tfrac{1}{2}\pi x\right) = (2/\pi)P_0(x) + \{(10\pi^2 - 120)/\pi^3\}P_2(x) = 0.6366 - 0.6871 P_2(x)$$

EXAMPLES

(1) Show that any quadratic $ax^2 + bx + c$ can be expressed in terms of P_0, P_1, and P_2. Find the values of a, b, c which minimize the integral

$$\int_{-1}^{1} \{\cos\left(\tfrac{1}{2}\pi x\right) - ax^2 - bx - c\}^2 \, \mathrm{d}x$$

(2) Obtain the approximation

$$\{(x + 1)/8\}^{\frac{1}{4}} = \tfrac{1}{3} + x/5 - P_2(x)/21 + P_3(x)/45$$

(3) Show that the coefficients in the expansion

$$|x| = a_0 P_0 + a_1 P_1 + a_2 P_2 + a_3 P_3 + a_4 P_4$$

are given by

$$a_0 = \tfrac{1}{2}, \qquad a_1 = 0, \qquad a_2 = \tfrac{5}{8}, \qquad a_3 = 0, \qquad a_4 = -\tfrac{3}{16}$$

(4) Obtain an approximation in the interval $(-1, 1)$ for the function exp x in the form

$$\exp x = a_0 P_0(x) + a_1 P_1(x) + a_2 P_2(x) + a_3 P_3(x)$$

(5) Show that the expansion of $\cos\left(\tfrac{1}{2}\pi x\right)$ in Legendre polynomials up to and including $P_4(x)$ is given by

$$\cos\left(\tfrac{1}{2}\pi x\right) = 0{\cdot}6366 P_0(x) - 0{\cdot}6871 P_2(x) + 0{\cdot}0518 P_4(x)$$

We now consider the evaluation for a particular value of x of a function $f(x)$ from its expansion in Legendre polynomials. Let the approximation, valid in $(-1, 1)$, be

$$f(x) = a_0 P_0 + a_1 P_1 + a_2 P_2 + a_3 P_3 + a_4 P_4$$

We form a sequence b_4, b_3, b_2, b_1, b_0 defined by

$$b_4 = a_4$$
$$b_3 = a_3 + \tfrac{7}{4}b_4 x$$
$$b_2 = a_2 + \tfrac{5}{3}b_3 x - \tfrac{3}{4}b_4$$
$$b_1 = a_1 + \tfrac{3}{2}b_2 x - \tfrac{2}{3}b_3$$
$$b_0 = a_0 + b_1 x - \tfrac{1}{2}b_2$$

Then the value of the approximation is b_0. This remarkable result is seen to be true if we multiply the first equation by $P_4(x)$, the second by $P_3(x)$, and so on, and then add the five equations together. By the recurrence relation

$$(n + 1)P_{n+1}(x) = (2n + 1)xP_n(x) - nP_{n-1}(x)$$

we have

$$b_4 P_4(x) = \tfrac{7}{4}b_4 x P_3(x) - \tfrac{3}{4}b_4 P_2(x)$$
$$b_3 P_3(x) = \tfrac{5}{3}b_3 x P_2(x) - \tfrac{2}{3}b_3 P_1(x)$$
$$b_2 P_2(x) = \tfrac{3}{2}b_2 x P_1(x) - \tfrac{1}{2}b_2 P_0(x)$$

and it follows that

$$b_0 P_0(x) = a_0 P_0(x) + a_1 P_1(x) + \cdots + a_4 P_4(x)$$

i.e., $f(x) = b_0$ approximately.

We have still to consider the accuracy of the approximation. Our guide to the error is the magnitude of the integral

$$\int_{-1}^{1} \{f(x) - (a_0 P_0 + a_1 P_1 + \cdots + a_4 P_4)\}^2 \, dx$$

This equals

$$\int_{-1}^{1} \{f(x)\}^2 \, dx - 2 \int_{-1}^{1} f(x)(a_0 P_0 + a_1 P_1 + \cdots + a_4 P_4) \, dx$$

$$+ \int_{-1}^{1} (a_0 P_0 + a_1 P_1 + \cdots + a_4 P_4)^2 \, dx$$

Since

$$\int_{-1}^{1} f(x)a_r P_r(x) \, dx = 2a_r^2/(2r + 1) \quad \text{and} \quad \int_{-1}^{1} P_m(x)P_n(x) \, dx = 0, \quad m \neq n$$

this gives

$$\int_{-1}^{1} \{f(x)\}^2 \, dx - \sum_{r=0}^{4} 2a_r^2/(2r + 1)$$

This is the integral of the square of the error over the interval $(-1, 1)$. If we include another term $a_5 P_5(x)$ in our approximation to $f(x)$, the integral would be reduced by $\frac{2}{11}a_5^2$. We calculate a_5 from the relation

$$a_5 = \tfrac{11}{2} \int_{-1}^{1} f(x)P_5(x) \, dx$$

If then $\frac{2}{11}a_5^2$ is negligible by our standard of accuracy, we conclude that the approximation would not be improved by the inclusion of more terms. Otherwise we add this term to the series and consider the magnitude of $\frac{2}{13}a_6^2$. This is not entirely satisfactory, as it gives us no guide to the size of the error at any particular point. When the coefficients in the expansion are decreasing fairly rapidly, as is commonly the case, it is reasonable to take the first non-zero neglected term to be a close approximation to the error. For example we have

$$\cos (\tfrac{1}{2}\pi x) = 0 \cdot 6366 P_0(x) - 0 \cdot 6871 P_2(x) + 0 \cdot 0518 P_4(x) + \cdots$$

If we use the approximation

$$\cos (\tfrac{1}{2}\pi x) = 0 \cdot 6366 P_0(x) - 0 \cdot 6871 P_2(x)$$

the error is given by $0 \cdot 0518 P_4(x)$. We can now judge how the error varies from point to point, by considering the curve

$$y = P_4(x) = (35x^4 - 30x^2 + 3)/8$$

The graph of $P_4(x)$ passes through the points $(1, 1), (-1, 1)$. It has a maximum at $(0, \tfrac{3}{8})$ and has minima at $(\pm\sqrt{\tfrac{3}{7}}, -\tfrac{3}{7})$. The graph crosses the x-axis at $x = \pm 0.340, \pm 0.861$. Calculation of the error confirms that it is given very closely by the value of $0.0518P_4(x)$. At $x = 0$, it is 0.02, at $x = \pm\sqrt{\frac{3}{7}}$ it is -0.02, and at $x = \pm 1$ it is 0.05; while at $x = \pm 0.340$ and ± 0.861 it is very small. It is not very satisfactory that the accuracy near the ends of the interval $(-1, 1)$ is much poorer than in the rest of the interval. This situation is characteristic of approximations by Legendre polynomials and is illustrated in Fig. 6.1.

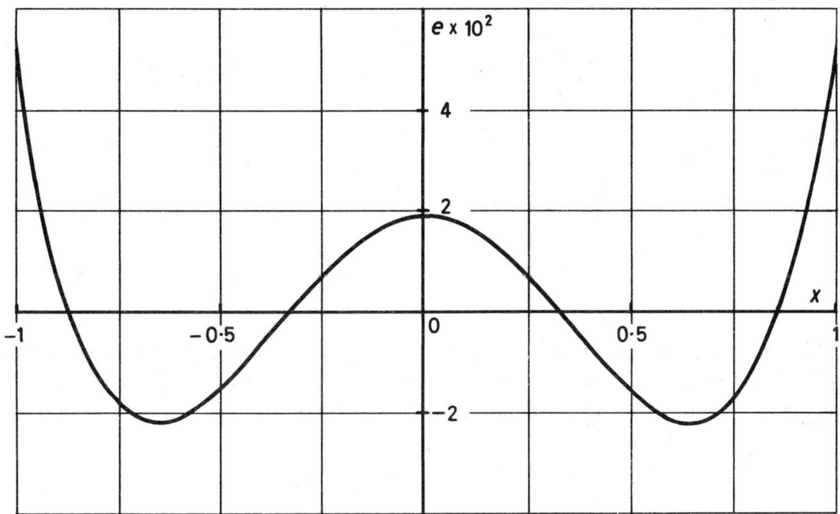

Fig. 6.1   The error e in the approximation $\cos\left(\tfrac{1}{2}\pi x\right) = 0.6366 - 0.6871P_2(x)$

A different method of obtaining this approximation is by using collocation at the zeros of $P_4(x)$. If x_1, x_2, x_3, and x_4 are the zeros of $P_4(x)$, the error term in the four-point Lagrange interpolation formula equals the product of $(x - x_1)(x - x_2)(x - x_3)(x - x_4)/4!$ and the value of the fourth derivative of $\cos\left(\tfrac{1}{2}\pi x\right)$ at some point ξ in $(-1, 1)$. The interpolation polynomial itself will contain no odd powers of x, since $\cos\left(\tfrac{1}{2}\pi x\right)$ is an even function. We can write it in the form $a_0P_0(x) + a_2P_2(x)$ and we can find the values of a_0 and a_2 from the fact that

$$\cos\left(\tfrac{1}{2}\pi x\right) = a_0P_0(x) + a_2P_2(x)$$

at the zeros of $P_4(x)$. We have therefore

(a) at $x = 0.340$, $0.8607 = a_0 - a_2(0.3266)$,
(b) at $x = 0.861$, $0.2166 = a_0 + a_2(0.6120)$.

Hence $a_0 = 0.637$, $a_2 = -0.686$, and therefore

$$\cos\left(\tfrac{1}{2}\pi x\right) = 0.637 - 0.686P_2(x) + \tfrac{1}{105}P_4(x)\{\pi/2\}^4 \cos\left(\tfrac{1}{2}\pi\xi\right)$$

When the approximation by a series of Legendre polynomials is required in an interval (a, b), we have merely to make the change of variable given by

$$2x = (b - a)t + (b + a)$$

so that the interval $a \leqslant x \leqslant b$ is transformed to the interval $-1 \leqslant t \leqslant 1$.

EXAMPLES

(1) Draw the graphs of $P_2(x)$, $P_3(x)$, $P_4(x)$, and $P_5(x)$ in $(-1, 1)$. Express these functions in terms of cosines of multiples of θ, where $x = \cos \theta$.

(2) Obtain the approximation

$$\sin^{-1}(x) = 1 \cdot 1781 P_1(x) + 0 \cdot 1718 P_3(x) + 0 \cdot 0675 P_5(x)$$

(3) Show that

(a) $(1 - x^2)^{\frac{1}{2}} = (\pi/2)\{\frac{1}{2} - \frac{5}{16}P_2(x) - \frac{9}{128}P_4(x) - \cdots\}$,

(b) $(1 - x^2)^{-\frac{1}{2}} = (\pi/2)\{1 + \frac{5}{4}P_2(x) + \frac{81}{64}P_4(x) + \cdots\}$.

(4) Obtain the expansion

$$\sin(\tfrac{1}{2}\pi x) = 1 \cdot 216 P_1(x) - 0 \cdot 225 P_3(x) + 0 \cdot 009 P_5(x)$$

Compare this result with that obtained by collocation at the zeros of $P_6(x)$.

Chebyshev polynomials

These polynomials are defined in the interval $(-1, 1)$ by $T_n(x) = \cos n\theta$, where $x = \cos \theta$. We have

$$T_0(x) = 1, \qquad T_1(x) = x$$
$$T_2(x) = 2x^2 - 1, \qquad T_3(x) = 4x^3 - 3x$$
$$T_4(x) = 8x^4 - 8x^2 + 1, \qquad T_5(x) = 16x^5 - 20x^3 + 5x$$

From the identity

$$\cos(n + 1)\theta + \cos(n - 1)\theta = 2 \cos \theta \cos n\theta$$

we have the recurrence relation

$$T_{n+1}(x) + T_{n-1}(x) = 2xT_n(x)$$

We can deduce at once that the coefficient of x^n in $T_n(x)$ is 2^{n-1}. Since $T_n(x) = \cos n\theta$, the zeros of $T_n(x)$ are at the points given by $n\theta = k\pi + \frac{1}{2}\pi$, i.e.,

$$x = \cos \theta = \cos\{(2k + 1)\pi/2n\}, \quad k = 0, 1, 2, \ldots, n - 1$$

The maximum values of $T_n(x)$ in $(-1, 1)$ come at the points

$$x = \cos(2k\pi/n), \quad (k \text{ an integer})$$

the maximum values all being equal to $+1$. The minimum values in $(-1, 1)$ come at the points

$$x = \cos\{(2k + 1)\pi/n\}, \quad (k \text{ an integer})$$

the minimum values all being equal to -1.

The usefulness of the Chebyshev polynomials is due to the fact that if $p(x)$ is any polynomial of degree n with the same leading coefficient 2^{n-1} as $T_n(x)$, then the largest value of $|p(x)|$ in the interval $(-1, 1)$ will be greater than 1, except when $p(x)$ and $T_n(x)$ are identically equal.

The Chebyshev polynomials form an orthogonal family with respect to integration over $(-1, 1)$ with weight function $1/\sqrt{(1 - x^2)}$, i.e.,

$$\int_{-1}^{1} \frac{T_m(x)T_n(x)}{\sqrt{(1 - x^2)}}\,dx = 0, \quad \text{if } m \neq n$$

If $m = n \neq 0$, we have

$$\int_{-1}^{1} \frac{T_n^2(x)}{\sqrt{(1 - x^2)}}\,dx = \tfrac{1}{2}\pi$$

The effect of the weight function is to give greater influence to the values of the function near $x = +1$ and near $x = -1$, and less influence to values near $x = 0$.

The expansion in the interval $(-1, 1)$ of a function $y(x)$ in terms of Chebyshev polynomials is usually written

$$y(x) = \tfrac{1}{2}a_0 T_0 + a_1 T_1 + a_2 T_2 + \cdots + a_n T_n$$

(The coefficient of T_0 is written $\tfrac{1}{2}a_0$ merely in order that the expression which we derive for a_r shall apply when $r = 0$.)

We can find the values of the coefficients $a_0, a_1, \ldots, a_n$ which will minimize the integral

$$\int_{-1}^{1} \{y - (\tfrac{1}{2}a_0 T_0 + a_1 T_1 + \cdots + a_n T_n)\}^2 \frac{dx}{\sqrt{(1 - x^2)}}$$

To do this we equate to zero the partial derivatives of this integral with respect to the coefficients, and we find

$$a_r = \frac{2}{\pi}\int_{-1}^{1} \frac{y(x)T_r(x)}{\sqrt{(1 - x^2)}}\,dx$$

The error in the approximation given by the series terminating in the term in $T_n(x)$ will be approximately equal to the first neglected term, i.e., to $a_{n+1}T_{n+1}$. Now this expression oscillates between the values $\pm a_{n+1}$ in the interval $(-1, 1)$, and vanishes at $(n + 1)$ points in the interval. This distribution of the error is more even than the distribution found in expansions in terms of Legendre polynomials, and for this reason expansions in terms of Chebyshev polynomials are preferred in subroutines for computers.

It is often not a simple matter to evaluate the integrals which give the values of the coefficients, and so it is usual to revert to collocation, using the points in the interval $(-1, 1)$ at which $T_{n+1}(x) = 0$, i.e., at the points $x_i = \cos \{(2i + 1)\pi/(2n + 2)\}$, $i = 0, 1, 2, \ldots, n$. In the Lagrange interpolation formula with $(n + 1)$ points

$$y(x) = \sum_{i=0}^{n} L_i(x)y_i + \pi(x)y^{(n+1)}(\xi)/(n + 1)!$$

we put $\pi(x) = T_{n+1}(x)/2^n$. Then we gain the advantage that the maximum value of the modulus of $\pi(x)$ is the least possible. Next we find constants $c_0, c_1, \ldots, c_n$ such that

$$\sum_{i=0}^{n} L_i(x)y_i = \tfrac{1}{2}c_0 + c_1 T_1 + c_2 T_2 + \cdots + c_n T_n$$

We require the difference

$$y(x) - \tfrac{1}{2}c_0 - c_1 T_1 - c_2 T_2 - \cdots - c_n T_n$$

to be zero at $x = x_i$, $i = 0, 1, 2, \ldots, n$. We make use now of another surprising property of the Chebyshev polynomials. The polynomials T_0, $T_1, \ldots, T_n$ form an orthogonal family with respect to summation over the zeros of $T_{n+1}(x)$. (Note that this is with weight function 1, not $(1 - x^2)^{-\frac{1}{2}}$.) This follows at once from the identity

$$\sum_i T_r(x_i)T_s(x_i) = \sum_i \cos (r\theta_i) \cos (s\theta_i)$$

$$= \tfrac{1}{2} \sum_i \{\cos (r - s)\theta_i + \cos (r + s)\theta_i\}$$

where

$$0 \leqslant r \leqslant n, \qquad 0 \leqslant s \leqslant n, \qquad \theta_i = \frac{2i + 1}{2n + 2} \pi \quad (i = 0, 1, 2, \ldots, n)$$

If $r \neq s$, this summation is zero.
If $r = s = 0$, the summation equals $(n + 1)$.
If $r = s \neq 0$, it equals $\tfrac{1}{2}(n + 1)$.
We multiply

$$y - \tfrac{1}{2}c_0 - c_1 T_1(x) - c_2 T_2(x) - \cdots - c_n T_n(x)$$

by $T_r(x)$ (where $0 \leqslant r \leqslant n$), and sum over the zeros of $T_{n+1}(x)$. We find

$$c_r = \frac{2}{n + 1} \sum_{i=0}^{n} T_r(x_i)y_i, \quad r = 0, 1, 2, \ldots, n$$

giving the coefficients in the expansion based on collocation at the zeros of $T_{n+1}(x)$.

To evaluate the series $(\tfrac{1}{2}c_0 + c_1 T_1 + c_2 T_2 + \cdots + c_n T_n)$ for a given value of x we form a sequence $b_n, b_{n-1}, \ldots, b_0$ as follows:

$$b_n = c_n$$
$$b_{n-1} = c_{n-1} + 2xb_n$$
$$b_{n-2} = c_{n-2} + 2xb_{n-1} - b_n$$
$$b_{n-3} = c_{n-3} + 2xb_{n-2} - b_{n-1}$$
$$\cdot \quad \cdot \quad \cdot \quad \cdot \quad \cdot \quad \cdot \quad \cdot \quad \cdot \quad \cdot$$
$$b_0 = c_0 + 2xb_1 - b_2$$

We multiply these equations by T_n, T_{n-1}, T_{n-2}, $T_{n-3}, \ldots, \frac{1}{2}T_0$ respectively and add. The recurrence relation gives

$$T_{r+1}(x) = 2xT_r(x) - T_{r-1}(x)$$

so that the coefficients of b_n, $b_{n-1}, \ldots, b_3$ and b_1 vanish and we obtain

$$\tfrac{1}{2}c_0 + c_1T_1 + c_2T_2 + \cdots + c_nT_n = \tfrac{1}{2}(b_0 - b_2)$$

We illustrate this by finding an approximation for the function $\cosh x$ in the form

$$\cosh x = \tfrac{1}{2}c_0 + c_2T_2(x) + c_4T_4(x)$$

Since $\cosh x$ is an even function there are no odd powers of x in the expansion. We will use the method of collocation, using the points at which $T_6(x)$ is zero, i.e., at the points $x = \pm\cos 15°$, $\pm\cos 45°$, $\pm\cos 75°$. By interpolation from six-figure tables we find the values

$$\cosh (\cos 15°) = 1\cdot503925, \qquad \cosh (\cos 45°) = 1\cdot260592$$
$$\cosh (\cos 75°) = 1\cdot033681$$

Then from the formula

$$c_r = \tfrac{1}{3} \sum_{i=0}^{5} T_r(x_i) \cosh (x_i)$$

$$c_0 = \tfrac{2}{3}(1\cdot503925 + 1\cdot260592 + 1\cdot033681) = 2\cdot532132$$
$$c_2 = (1/\sqrt{3})(1\cdot503925 - 1\cdot033681) = 0\cdot271495$$
$$c_4 = \tfrac{1}{3}(1\cdot503925 - 2\cdot521184 + 1\cdot033681) = 0\cdot005474$$

This gives the result

$$\cosh x = 1\cdot266066 + 0\cdot271495T_2(x) + 0\cdot005474T_4(x)$$

We now evaluate this series at $x = 0\cdot75$ by means of the recurrence scheme above. We find

$$b_4 = 0\cdot005474, \qquad b_3 = 0\cdot008211$$
$$b_2 = 0\cdot278338, \qquad b_1 = 0\cdot409295$$
$$b_0 = 2\cdot867738$$

Then
$$\cosh (0\cdot75) = \tfrac{1}{2}(b_0 - b_2) = 1\cdot294700$$

Since the correct value is 1·294683, we are in error by 17 units. Compare this with the result given by the truncated Taylor series

$$\cosh x = 1 + x^2/2 + x^4/24$$

At $x = 0.75$ this gives 1·294434, with an error of 249 units. The accuracy of this approximation is very good near the origin, but gets progressively worse as x increases, the error at $x = 1$ being 1414 units. Contrast this with the accuracy of the Chebyshev series above, in which the error never exceeds 46 units in the range $-1 \leqslant x \leqslant 1$, but in which the error is very nearly 46 units at the points $x = 0,\ \pm\frac{1}{2},\ \pm\frac{1}{2}\sqrt{3},\ \pm1$. The error curves for these approximations are shown in Fig. 6.2.

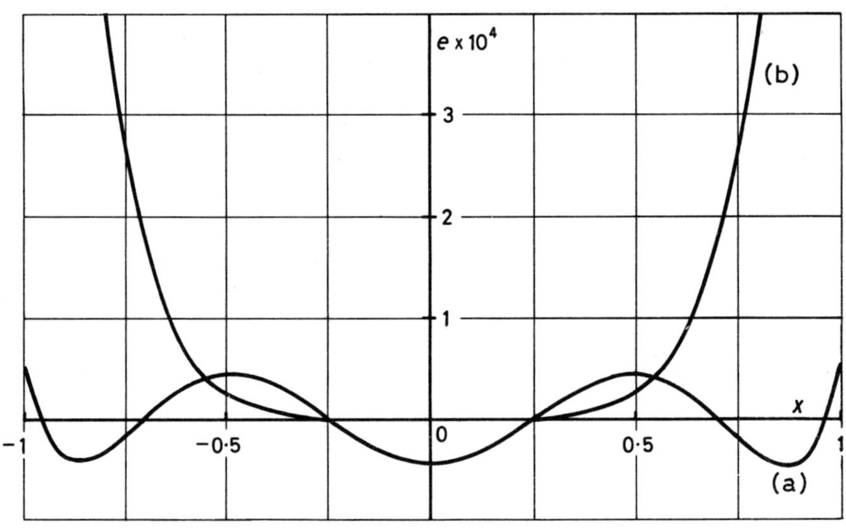

Fig. 6.2   The error e in the approximations
(a) $\cosh x = 1\cdot266066 + 0\cdot271495T_2(x) + 0\cdot005474T_4(x)$
(b) $\cosh x = 1 + x^2/2 + x^4/24$

EXAMPLES

(1) Draw the graphs of the functions $T_2(x)$, $T_3(x)$, $T_4(x)$, and $T_5(x)$ in the range $(-1, 1)$.

(2) Express x^3, x^4, and x^5 in terms of Chebyshev polynomials.

(3) Obtain by the method of least squares the expansion

$$|x| = \frac{4}{\pi}\left\{\frac{1}{2} + \frac{1}{3}T_2(x) - \frac{1}{15}T_4(x) + \frac{1}{35}T_6(x) - \frac{1}{63}T_8(x) + \cdots\right\}$$

(4) By collocation at the zeros of $T_{12}(x)$, obtain the approximation

$$|x| = 0\cdot638 + 0\cdot422T_2(x) - 0\cdot081T_4(x) + 0\cdot032T_6(x)$$
$$-0\cdot015T_8(x) + 0\cdot006T_{10}(x)$$

(5) Obtain the expansions

(a) $\sin(\tfrac{1}{2}\pi x) = 1\!\cdot\!1336T_1(x) - 0\!\cdot\!1381T_3(x) + 0\!\cdot\!0045T_5(x)$
(b) $\cos(\tfrac{1}{2}\pi x) = 0\!\cdot\!4720 - 0\!\cdot\!4994T_2(x) + 0\!\cdot\!0280T_4(x)$

Find the error in the second expansion at the zeros of $T_6(x)$.
(6) Show that

$$\sinh^{-1} x = x(0\!\cdot\!9359 - 0\!\cdot\!0588T_2 + 0\!\cdot\!0047T_4$$
$$- 0\!\cdot\!0005T_6 + 0\!\cdot\!0001T_8 - \cdots)$$

Use the expansion to calculate the value of $\sinh^{-1}(\tfrac{1}{2})$ by means of a recurrence relation. Compare the result with the exact value, $\log_e \tfrac{1}{2}(\sqrt{5} + 1)$.
(7) Obtain the expansion

$$\tan(\tfrac{1}{4}\pi x) = x(0\!\cdot\!88507 + 0\!\cdot\!10675T_2 + 0\!\cdot\!00759T_4$$
$$+ 0\!\cdot\!00054T_6 + 0\!\cdot\!00003T_8)$$

Evaluate the expansion for $x = 1$, $x = -1$, and $x = \tfrac{1}{3}$.

Integration and differentiation of Chebyshev series

If y is a polynomial in x of degree n we can express it in the form

$$y = \tfrac{1}{2}a_0 + a_1T_1 + a_2T_2 + \cdots + a_nT_n$$

Its derivative will be of degree $n - 1$, so that we can write

$$y' = \tfrac{1}{2}b_0 + b_1T_1 + b_2T_2 + \cdots + b_{n-1}T_{n-1}$$

In order to find the relation between the two sets of coefficients we need to integrate $T_r(x)$.

We have $\int T_1(x)\,\mathrm{d}x = \tfrac{1}{4}T_2(x) + $ a constant k.

$$\int T_2(x)\,\mathrm{d}x = -\int \cos 2\theta \sin \theta\,\mathrm{d}\theta = \tfrac{1}{2}\int (\sin \theta - \sin 3\theta)\,\mathrm{d}\theta$$
$$= T_3(x)/6 - T_1(x)/2 + k$$

$$\int T_3(x)\,\mathrm{d}x = T_4(x)/8 - T_2(x)/4 + k$$

In general, for $r > 1$,

$$\int T_r(x)\,\mathrm{d}x = T_{r+1}(x)/(2r + 2) - T_{r-1}(x)/(2r - 2) + k$$

We can now integrate the series above for y'. This gives

$$y = \tfrac{1}{2}(b_0 - b_2)T_1 + \tfrac{1}{4}(b_1 - b_3)T_2 + \tfrac{1}{6}(b_2 - b_4)T_3 + \cdots$$
$$+ (b_{n-3} - b_{n-1})T_{n-2}/(2n - 4) + b_{n-2}T_{n-1}/(2n - 2) + b_{n-1}T_n/2n + k.$$

By comparing the two series for y we obtain the relations

$$b_0 - b_2 = 2a_1, \qquad b_{n-3} - b_{n-1} = (2n - 4)a_{n-2}$$
$$b_1 - b_3 = 4a_2, \qquad b_{n-2} = (2n - 2)a_{n-1}$$
$$b_2 - b_4 = 6a_3, \qquad b_{n-1} = 2na_n$$

In the same way we obtain relations between the coefficients in the series for y' and those in the series

$$y'' = \tfrac{1}{2}c_0 + c_1 T_1 + c_2 T_2 + \cdots + c_{n-2} T_{n-2}$$

These are $\qquad c_{r-1} - c_{r+1} = 2rb_r, \quad r = 1, 2, \ldots, n - 3$

and $\qquad\qquad c_{r-1} = 2rb_r, \quad r = n - 2, n - 1$

An alternative approach can be made by considering the derivatives of the Chebyshev polynomials. We have

$$DT_1(x) = 1$$
$$DT_2(x) = 4x = 2(2T_1)$$
$$DT_3(x) = 12x^2 - 3 = 3(2T_2 + 1)$$
$$DT_4(x) = 32x^3 - 16x = 4(2T_3 + 2T_1)$$

In general, from

$$T_{r+1}/(r + 1) = 2\int T_r(x)\,dx + T_{r-1}/(r - 1) + k$$

we find

$$\begin{aligned}
DT_{r+1}/(r + 1) &= 2T_r + DT_{r-1}/(r - 1) \\
&= 2T_r + 2T_{r-2} + DT_{r-3}/(r - 3) \\
&= 2T_r + 2T_{r-2} + 2T_{r-4} + DT_{r-5}/(r - 5) \quad (r > 5)
\end{aligned}$$

Hence for n even

$$DT_n = n(2T_{n-1} + 2T_{n-3} + \cdots + 2T_1)$$

and for n odd

$$DT_n = n(2T_{n-1} + 2T_{n-3} + \cdots + 2T_2 + 1)$$

By differentiating the series for y we find

$$\begin{aligned}
b_0 &= 2(a_1 + 3a_3 + 5a_5 + 7a_7 + \cdots) \\
b_1 &= 2(2a_2 + 4a_4 + 6a_6 + 8a_8 + \cdots) \\
b_2 &= 2(3a_3 + 5a_5 + 7a_7 + 9a_9 + \cdots) \\
b_r &= 2\{(r + 1)a_{r+1} + (r + 3)a_{r+3} + \cdots\}
\end{aligned}$$

This is in agreement with the relation $b_{r-1} - b_{r+1} = 2ra_r$.

Whichever method we use, there is a magnification of round-off error when we calculate the coefficients b_r from the coefficients a_r.

Application to differential equations

By the recurrence relation

$$2xT_r = T_{r+1} + T_{r-1} \quad (r > 0)$$

we have
$$4x^2 T_r = T_{r+2} + 2T_r + T_{r-2} \quad (r > 1)$$

and
$$8x^3 T_r = T_{r+3} + 3T_{r+1} + 3T_{r-1} + T_{r-3} \quad (r > 2)$$

We can use relations of this type to express the product of a polynomial in x with any of the series above for y, y', and y'' as a series of Chebyshev polynomials. In particular, we have

$$2xy = a_1 + (a_0 + a_2)T_1 + (a_1 + a_3)T_2 + \cdots$$
$$+ (a_{n-2} + a_n)T_{n-1} + a_{n-1}T_n + a_nT_{n+1}$$
$$4x^2y = (a_0 + a_2) + (3a_1 + a_3)T_1 + (a_0 + 2a_2 + a_4)T_2 + \cdots$$
$$+ (a_{n-2} + 2a_n)T_n + (a_{n-1})T_{n+1} + a_nT_{n+2}$$

We can apply these formulae to solve linear differential equations in which the coefficients are polynomials in x. We begin by considering the equation $y' = y$, $y(0) = 1$, which has the solution $y = e^x$. We assume an expansion

$$y' = \tfrac{1}{2}b_0 + b_1T_1 + b_2T_2 + b_3T_3 + b_4T_4$$

and by integrating this we find

$$y = \tfrac{1}{2}(b_0 - b_2)T_1 + \tfrac{1}{4}(b_1 - b_3)T_2 + \tfrac{1}{6}(b_2 - b_4)T_3 + \tfrac{1}{8}b_3T_4 + \tfrac{1}{10}b_4T_5 + k.$$

Hence if we make $b_0 - b_2 = 2b_1$, $b_1 - b_3 = 4b_2$, $b_2 - b_4 = 6b_3$, $b_3 = 8b_4$, and $k = \tfrac{1}{2}b_0$, we have $y - y' = b_4T_5/10$. These relations give $b_0 = 457b_4$, $b_1 = 204b_4$, $b_2 = 49b_4$. We also have from the condition $y(0) = 1$

$$\tfrac{1}{2}b_0 - b_2 + b_4 = 1$$

and so $b_4 = 2/361$. Our series become

$$y = (457 + 408T_1 + 98T_2 + 16T_3 + 2T_4 + \tfrac{1}{5}T_5)/361$$
$$y' = (457 + 408T_1 + 98T_2 + 16T_3 + 2T_4)/361$$

which satisfy exactly the equation

$$y - y' = T_5/1805, \qquad y(0) = 1$$

We can compare this with the approximation

$$y = 1 + x + x^2/2! + x^3/3! + x^4/4! + x^5/5!$$

which satisfies exactly the equation

$$y - y' = x^5/120, \quad y(0) = 1$$

Next we consider the equation $y' = 1 - 2xy$, $y(0) = 0$. Since all the even-order derivatives of y are zero at $x = 0$, we can put

$$y = a_1T_1 + a_3T_3 + a_5T_5 + a_7T_7$$

By using the relation $2xT_n = T_{n+1} + T_{n-1}$, we obtain

$$1 - 2xy = (1 - a_1) - (a_1 + a_3)T_2 - (a_3 + a_5)T_4 - (a_5 + a_7)T_6 - a_7T_8$$

Since this is a polynomial of degree eight we cannot equate it to y' which is of degree six. We integrate the series

$$y' = \tfrac{1}{2}b_0 + b_2T_2 + b_4T_4 + b_6T_6$$

and obtain

$$y = \tfrac{1}{2}(b_0 - b_2)T_1 + \tfrac{1}{6}(b_2 - b_4)T_3 + \tfrac{1}{10}(b_4 - b_6)T_5 + \tfrac{1}{14}b_6T_7$$

We compare this with the original series for y and find

$$b_0 - b_2 = 2a_1, \qquad b_2 - b_4 = 6a_3, \qquad b_4 - b_6 = 10a_5, \qquad b_6 = 14a_7$$

Since we have eight unknowns we need four more equations, and these we obtain by equating the first four coefficients in the series for y' and $(1 - 2xy)$:

$$\tfrac{1}{2}b_0 = 1 - a_1, \quad b_2 = -a_1 - a_3, \quad b_4 = -a_3 - a_5, \quad b_6 = -a_5 - a_7$$

From these eight equations we find

$$3a_1 = 2 + a_3, \quad 6a_3 = a_5 - a_1, \quad 10a_5 = a_7 - a_3, \quad 15a_7 = -a_5$$

The solution of these equations gives the result

$$y = (921T_1 - 151T_3 + 15T_5 - T_7)/1457$$
$$y' = (536 - 770T_2 + 136T_4 - 14T_6)/1457$$

while $y' = 1 - 2xy - T_8/1457$.

If we include more terms in the series for y, the system of equations becomes $3a_1 = 2 + a_3$, $2na_n = a_{n+2} - a_{n-2}$ for n odd, $n > 1$. We can draw up a computer program for the solution by Gauss–Seidel iteration. Taking initial values $a_1 = \tfrac{2}{3}$, $a_n = 0$, $n > 1$, we substitute these values in the right-hand side of each equation to find new values on the left-hand. We do this repeatedly until the iteration has converged.

To solve the equation $(1 + x^2)y' = 1$, $y(0) = 0$, we put

$$y' = \tfrac{1}{2}b_0 + b_2T_2 + b_4T_4 + b_6T_6$$

Then

$$4x^2y' = (b_0 + b_2) + (b_0 + 2b_2 + b_4)T_2 + (b_2 + 2b_4 + b_6)T_4$$
$$+ (b_4 + 2b_6)T_6 + b_6T_8$$

We need four equations in order to find the four unknowns and so we equate to zero the coefficients of T_0, T_2, T_4, and T_6 in $(1 + x^2)y' - 1$. This gives

$$3b_0 + b_2 = 4, \qquad b_0 + 6b_2 + b_4 = 0$$
$$b_2 + 6b_4 + b_6 = 0, \qquad b_4 + 6b_6 = 0$$

The solution of these equations gives the series

$$y' = (408 - 140T_2 + 24T_4 - 4T_6)/577$$

and hence

$$y = (478T_1 - 82T_3/3 + 14T_5/5 - 2T_7/7)/577$$

This polynomial satisfies the equation

$$(1 + x^2)y' = 1 - T_8/577, \qquad y(0) = 0$$

Hence

$$\tan^{-1} x - y = \int_0^x \{T_8/577(1 + x^2)\}\, dx$$

For $-1 \leqslant x \leqslant 1$, the right-hand side is less than $\int_0^a T_8\, dx/577$ where $a = \cos(7\pi/16)$. This again is less than $0{\cdot}00022$, so that the error in the approximation

$$\tan^{-1} x = (478T_1 - 82T_3/3 + 14T_5/5 - 2T_7)/577$$

is less than $2{\cdot}2 \times 10^{-4}$ in $(-1, 1)$.

Finally we consider the second-order boundary-value problem

$$(1 + x)y'' + xy' - y = 0, \qquad y(1) = 1, \qquad y(-1) = 1$$

If

$$y = \tfrac{1}{2}a_0 + a_1 T_1 + a_2 T_2 + a_3 T_3$$

the boundary conditions require

$$\tfrac{1}{2}a_0 + a_1 + a_2 + a_3 = 1, \qquad \tfrac{1}{2}a_0 - a_1 + a_2 - a_3 = 1$$

i.e.,

$$a_0 + 2a_2 = 2, \qquad\qquad a_1 + a_3 = 0$$

By integrating $y' = \tfrac{1}{2}b_0 + b_1 T_1 + b_2 T_2$ and $y'' = \tfrac{1}{2}c_0 + c_1 T_1$, and comparing coefficients we find

$$2a_1 = b_0 - b_2, \qquad 4a_2 = b_1, \qquad 6a_3 = b_2, \qquad 2b_1 = c_0, \qquad 4b_2 = c_1$$

Using the relations

$$2xy' = b_1 + (b_0 + b_2)T_1 + b_1 T_2 + b_2 T_3$$
$$2xy'' = c_1 + c_0 T_1 + c_1 T_2$$

we find

$$(1 + x)y'' + xy' - y = \tfrac{1}{2}(c_0 + c_1 + b_1 - a_0)$$
$$+ \tfrac{1}{2}(c_0 + 2c_1 + b_0 + b_2 - 2a_1)T_1$$
$$+ \tfrac{1}{2}(c_1 + b_1 - 2a_2)T_2 + \tfrac{1}{2}(b_2 - 2a_3)T_3$$

We have nine unknowns and we have already seven equations which they must satisfy. We need only two more equations, for which we take

$$c_0 + c_1 + b_1 - a_0 = 0$$
$$c_0 + 2c_1 + b_0 + b_2 - 2a_1 = 0$$

The solution of the nine equations leads to the result

$$y = (66 + 2T_1 + 15T_2 - 2T_3)/81$$
$$y' = (-4 + 60T_1 - 12T_2)/81$$
$$y'' = (60 - 48T_1)/81$$

Our series satisfy exactly the equation

$$(1 + x)y'' + xy' - y = -(9T_2 + 4T_3)/81, \quad y(1) = 1, y(-1) = 1$$

EXAMPLES

(1) Show that $y = T_6(x)$ is the solution of the equation

$$(1 - x^2)y'' - xy' + 36y = 0, \quad y(1) = 1, \quad y(-1) = 1$$

(2) Show that the solution of the equation

$$(2x^2 - 1)y' - 8xy = 12x - 8x^3, \quad y(0) = 0$$

is $y = T_2(x) + T_4(x)$.

(3) Show that the solution of the equation $y' + y = x$, $y(0) = 1$ is given approximately by

$$y = 1{\cdot}532 - 1{\cdot}260T_1 + 0{\cdot}543T_2 - 0{\cdot}088T_3 + 0{\cdot}012T_4$$

(4) Obtain the expansion of the function $(\tan^{-1} x)/x$ in the form

$$0{\cdot}8814 - 0{\cdot}1059T_2 + 0{\cdot}0111T_4 - 0{\cdot}0014T_6 + 0{\cdot}0002T_8$$

(5) The Bessel function $J_0(x)$ satisfies the differential equation

$$xy'' + y' + xy = 0, \quad y(0) = 1, y'(0) = 0$$

Show that

$$J_0(x) = 0{\cdot}8807 - 0{\cdot}1174T_2 + 0{\cdot}0019T_4 - \cdots$$

and hence evaluate $J_0(x)$ for $x = 0(0{\cdot}2)1$.

(6) The Bessel function $J_0(4x)$ satisfies the differential equation

$$xy'' + y' + 16xy = 0, \quad y(0) = 1, y'(0) = 0$$

Deduce the expansion

$$J_0(4x) = 0{\cdot}0501 - 0{\cdot}6653T_2 + 0{\cdot}2490T_4$$
$$-0{\cdot}0332T_6 + 0{\cdot}0023T_8 - 0{\cdot}0001T_{10}$$

Hence evaluate $J_0(x)$ for $x = 1, 2, 3, 4$.

(7) Use the expansion of $J_0(x)$ in a series of Chebyshev polynomials to express

$$f(x) = \int_0^x J_0(x)\,dx$$

as a series of Chebyshev polynomials, and hence evaluate $f(x)$ for $x = 0(0·25)1$.

(8) If $y = \frac{1}{2}a_0 + a_1 T_1 + a_2 T_2 + \cdots$, show that

$$xy = \frac{1}{2}a_1 + \frac{1}{2}(a_0 + a_2)T_1 + \frac{1}{2}(a_1 + a_3)T_2 + \cdots$$
$$x^2 y = \frac{1}{4}(a_0 + a_2) + \frac{1}{4}(3a_1 + a_3)T_1 + \frac{1}{4}(a_0 + 2a_2 + a_4)T_2 + \cdots$$

Harmonic analysis

For dealing with experimental results, one common method is to obtain a Fourier series as an approximation to the required function. To do this, we have to calculate the coefficients in a series of sines and cosines of multiples of x. If the function $f(x)$ is periodic with period 2π, i.e., if $f(x + 2\pi) = f(x)$, we can represent $f(x)$ by the series

$$\frac{1}{2}a_0 + a_1 \cos x + a_2 \cos 2x + \cdots + a_n \cos nx$$
$$+ b_1 \sin x + b_2 \sin 2x + \cdots + b_n \sin nx$$

If the period is not equal to 2π, we can always make a suitable change of variable. If $f(x)$ is not periodic, and is defined only in the range $0 \leqslant x \leqslant 2a$ we can represent it by the series

$$\frac{1}{2}a_0 + a_1 \cos (\pi x/a) + a_2 \cos (2\pi x/a) + \cdots + a_n \cos (n\pi x/a)$$
$$+ b_1 \sin (\pi x/a) + b_2 \sin (2\pi x/a) + \cdots + b_n \sin (n\pi x/a)$$

This is especially useful when the derivatives of the function are continuous and satisfy the relation $f^{(n)}(0) = f^{(n)}(2a)$.

The usefulness of this series representation springs from two properties of sines and cosines. First, their sums, products, integrals, and derivatives are all expressible as sines and cosines. Second, they are orthogonal both for integration over the interval $(0, 2\pi)$, and for summation over a suitable set of points in this range.

We consider first the use of integration. If p and q are integers

$$\int_0^{2\pi} \cos px \cos qx \, dx = 0, \quad p \neq q$$
$$= \pi, \quad p = q \neq 0$$
$$= 2\pi, \quad p = q = 0$$

$$\int_0^{2\pi} \cos px \sin qx \, dx = 0, \quad \text{all } p, q$$

$$\int_0^{2\pi} \sin px \sin qx \, dx = 0, \quad p \neq q$$
$$= \pi, \quad p = q \neq 0$$
$$= 0, \quad p = q = 0$$

We now apply the method of least squares to determine the coefficients in the series. We wish to minimize the integral

$$I = \int_0^{2\pi} \left\{ f(x) - \tfrac{1}{2}a_0 - \sum_{k=1}^{n} (a_k \cos kx + b_k \sin kx) \right\}^2 dx$$

By differentiating partially with respect to each coefficient in turn and equating the derivatives to zero, we find

$$\pi a_k = \int_0^{2\pi} f(x) \cos kx\, dx, \quad k = 0, 1, 2, \ldots, n$$

$$\pi b_k = \int_0^{2\pi} f(x) \sin kx\, dx, \quad k = 1, 2, 3, \ldots, n$$

The coefficients are independent of each other, so that we can extend the series without having to recalculate all the coefficients. Using the orthogonality of the sines and cosines, we can show that the minimum value of the integral I above is

$$\int_0^{2\pi} \{f(x)\}^2\, dx - \pi\left\{ \tfrac{1}{2}a_0^2 + \sum_{k=1}^{n} (a_k^2 + b_k^2) \right\}$$

As n is increased, this minimum value will tend to zero, if $f(x)$ is continuous and differentiable in the range (except possibly for a finite number of points). The sum of the Fourier series will approach the value of $f(x)$ at each point at which $f(x)$ is continuous.

Next we consider the method of summation. The orthogonality properties for summation over the set of points $x_r = \pi r/n (r = 0, 1, 2, \ldots, 2n - 1)$ are as follows. If p and q are zero or are positive integers less than $2n$, we have

$$\sum_{r=0}^{2n-1} \cos px_r \cos qx_r = \sum_{r=0}^{2n-1} \sin px_r \sin qx_r = 0, \quad p \neq q$$

Each of these sums equals n if $p = q \neq 0$ or n. Clearly if $p = q = 0$ or $p = q = n$, the cosine sum equals $2n$ and the sine sum equals 0. Also we have

$$\sum_{r=0}^{2n-1} \cos px_r \sin qx_r = 0$$

We now minimize the sum

$$S = \sum_{r=0}^{2n-1} \left\{ f(x_r) - \tfrac{1}{2}A_0 - \sum_{k=1}^{n} (A_k \cos kx_r + B_k \sin kx_r) \right\}^2$$

By the same method as before, we obtain the following expressions for the coefficients

$$nA_k = \sum_{r=0}^{2n-1} f(x_r) \cos kx_r, \quad k = 0, 1, 2, \ldots, n-1$$

$$2nA_n = \sum_{r=0}^{2n-1} f(x_r) \cos nx_r$$

$$nB_k = \sum_{r=0}^{2n-1} f(x_r) \sin kx_r, \quad k = 1, 2, \ldots, n-1$$

Since $\sin nx_r = 0$ for each r, we have $B_n = 0$. Our approximation is now

$$\tfrac{1}{2}A_0 + A_1 \cos x + A_2 \cos 2x + \cdots + A_n \cos nx$$
$$+ B_1 \sin x + B_2 \sin 2x + \cdots + B_{n-1} \sin (n-1)x$$

The sum of this series will equal $f(x)$ at each point $x_r = \pi r/n$, ($r = 0, 1, 2, \ldots, 2n - 1$), so that the minimum value of the sum S above is zero. The method of collocation will give the same result, as can be seen by considering the set of simultaneous equations

$$f(x_r) = \tfrac{1}{2}A_0 + \sum_{k=1}^{n} A_k \cos kx_r + \sum_{k=1}^{n-1} B_k \sin kx_r, \quad r = 0, 1, 2, \ldots, 2n - 1$$

We regard $A_0, A_1, \ldots, A_n, B_1, \ldots, B_{n-1}$ as the unknowns, and we have $2n$ linear equations to solve for $2n$ unknowns. If we multiply the rth equation by $\cos kx_{r-1}$ and add, we obtain the equation given above for A_k. If we multiply instead by $\sin kx_{r-1}$, we obtain the equation for B_k.

There is an interesting relation between the coefficients a_k, b_k found by integration and the coefficients A_k, B_k found by summation. The integrals for a_k and b_k will have to be evaluated numerically when the data are experimental measurements, or when the analytic integration is too difficult. Using the trapezium rule with $h = \pi/n$, we find, since $f(0) = f(2\pi)$,

$$n \int_0^{2\pi} f(x) \cos kx \, dx = \pi \sum_{r=1}^{2n-1} f(x_r) \cos kx_r, \quad k = 0, 1, 2, \ldots, n-1$$

$$2n \int_0^{2\pi} f(x) \cos nx \, dx = \pi \sum_{r=0}^{2n-1} f(x_r) \cos nx_r$$

$$n \int_0^{2\pi} f(x) \sin kx \, dx = \pi \sum_{r=0}^{2n-1} f(x_r) \sin kx_r, \quad k = 0, 1, 2, \ldots, n-1$$

Thus the coefficients A_k, B_k equal the approximate values found for a_k, b_k. The accuracy of these approximations diminishes as k increases. Consider for example the case in which $2n = 12$ and $k = 4$. We find

$$B_4 = \frac{\sqrt{3}}{12}\{f(x_1) - f(x_2) + f(x_4) - f(x_5) + f(x_7)$$
$$-f(x_8) + f(x_{10}) - f(x_{11})\}$$

This is the coefficient of $\sin 4x$, which completes four cycles in the range $(0, 2\pi)$, and in each cycle we are using only two values of $f(x)$ to determine the coefficient B_4. For this reason, when the method of summation is used with $2n = 12$, we rely on the accuracy of the coefficients A_0, A_1, A_2, B_1, and B_2, but not on the accuracy of the remaining coefficients. If they are found not to be negligible, then the value of $2n$ should be increased.

There is another point to be borne in mind. When $2n = 12$, the values of $\cos 7x_r$ and $\cos 5x_r$ are equal for all values of the integer r. Thus a component of $\cos 7x$ will appear in the result found by summation as a component of $\cos 5x$. In the same way, $\cos 8x_r$ appears as $\cos 4x_r$ and $\sin 7x_r$ appears as $-\sin 5x_r$. This is known as 'aliasing', and the frequencies are said to 'fold'.

For the calculation of the coefficients when $2n = 12$ the following notation is convenient, with y_r denoting the value of $f(x_r)$.

$$\alpha_0 = y_0 + y_6, \qquad \alpha_1 = y_1 + y_7, \qquad \alpha_2 = y_2 + y_8$$
$$\alpha_3 = y_3 + y_9, \qquad \alpha_4 = y_4 + y_{10}, \qquad \alpha_5 = y_5 + y_{11}$$
$$\beta_0 = y_0 - y_6, \qquad \beta_1 = y_1 - y_7, \qquad \beta_2 = y_2 - y_8$$
$$\beta_3 = y_3 - y_9, \qquad \beta_4 = y_4 - y_{10}, \qquad \beta_5 = y_5 - y_{11}$$

$$6A_0 = \alpha_0 + \alpha_1 + \alpha_2 + \alpha_3 + \alpha_4 + \alpha_5$$
$$6A_1 = \beta_0 + \tfrac{1}{2}\sqrt{3}(\beta_1 - \beta_5) + \tfrac{1}{2}(\beta_2 - \beta_4)$$
$$6A_2 = (\alpha_0 - \alpha_3) + \tfrac{1}{2}(\alpha_1 - \alpha_2 - \alpha_4 + \alpha_5)$$
$$6A_3 = \beta_0 - \beta_2 + \beta_4$$
$$6A_4 = (\alpha_0 + \alpha_3) - \tfrac{1}{2}(\alpha_1 + \alpha_2 + \alpha_4 + \alpha_5)$$
$$6A_5 = \beta_0 - \tfrac{1}{2}\sqrt{3}(\beta_1 - \beta_5) + \tfrac{1}{2}(\beta_2 - \beta_4)$$
$$6A_6 = \tfrac{1}{2}(\alpha_0 - \alpha_1 + \alpha_2 - \alpha_3 + \alpha_4 - \alpha_5)$$

$$6B_1 = \tfrac{1}{2}(\beta_1 + \beta_5) + \beta_3 + \tfrac{1}{2}\sqrt{3}(\beta_2 + \beta_4)$$
$$6B_2 = \tfrac{1}{2}\sqrt{3}(\alpha_1 + \alpha_2 - \alpha_4 - \alpha_5)$$
$$6B_3 = \beta_1 - \beta_3 + \beta_5$$
$$6B_4 = \tfrac{1}{2}\sqrt{3}(\alpha_1 - \alpha_2 + \alpha_4 - \alpha_5)$$
$$6B_5 = \tfrac{1}{2}(\beta_1 + \beta_5) + \beta_3 - \tfrac{1}{2}\sqrt{3}(\beta_2 + \beta_4)$$

For checking, we can employ the identities

(a) $$y_0 = \tfrac{1}{2}A_0 + A_1 + A_2 + \ldots + A_6$$

(b) $$y_1 - y_{11} = B_1 + \sqrt{3}B_2 + 2B_3 + \sqrt{3}B_4 + B_5$$

We have used the range $(0, 2\pi)$ exclusively so far, but for some purposes the range $(-\pi, \pi)$ has advantages. This is so when the function is even, i.e. $f(-x) = f(x)$, so that all the coefficients $B_1, B_2, \ldots$ are clearly zero, or when the function is odd, i.e., $f(-x) = -f(x)$, in which case $A_0, A_1, A_2, \ldots$ are zero. Our formulae are easily adapted to this range.

It is not surprising that there is a close connection between Fourier cosine series and the Chebyshev polynomials. Consider a function $F(x)$ defined in the interval $-1 \leqslant x \leqslant 1$. Let $f(\theta)$ equal $F(\cos \theta)$ so that $f(\theta)$ is defined in the interval $0 \leqslant \theta \leqslant \pi$, and extend the range of definition of $f(\theta)$ by putting $f(2\pi - \theta) = f(\theta)$. We now obtain the expansion of $f(\theta)$ in a series of cosines and sines, and we find that the coefficients $B_1, B_2, \ldots, B_{n-1}$ are all zero. Then from

$$f(\theta) = \tfrac{1}{2}A_0 + A_1 \cos \theta + A_2 \cos 2\theta + \cdots + A_n \cos n\theta$$

we obtain the expansion

$$F(x) = \tfrac{1}{2}A_0 + A_1 T_1(x) + A_2 T_2(x) + \cdots + A_n T_n(x)$$

the coefficients being defined as above by summation over the points $x_r = \cos (\pi r/n)$. In using this method for obtaining the expansion of $F(x)$ as a series of Chebyshev polynomials, we must be certain that any component of $T_{n+1}(x), T_{n+2}(x), \ldots$ in $F(x)$ is negligible. It can be seen that this method is equivalent to collocation at the points $x_r = \cos (\pi r/n)$, whereas a similar method considered earlier used collocation at the points

$$x_i = \cos \{(2i + 1)\pi/(2n + 2)\}$$

EXAMPLES

(1) The values of the function y are measured experimentally for $x = 0, h, 2h, \ldots, 12h$, and are found to be as follows.

$$
\begin{array}{ll}
y_0 = y_{12} = 3.25, & y_6 = 1.25 \\
y_1 = 4.12, & y_7 = 2.38 \\
y_2 = 3.75, & y_8 = 2.75 \\
y_3 = 2.25, & y_9 = 2.25 \\
y_4 = 0.75, & y_{10} = 1.75 \\
y_5 = 0.38, & y_{11} = 2.12
\end{array}
$$

Find an approximation for y by putting $\theta = \pi x/6h$.

(2) Show from the following values of $y(x)$, at the points $x = 0(\pi/6)2\pi$, that y can be expressed as the sum of a sine and cosine in the range $0 \leqslant x \leqslant 2\pi$.

$$
\begin{array}{lll}
y_0 = y_6 = y_{12} = 2.4, & y_1 = y_5 = 2.8, & y_3 = -4.0 \\
y_2 = y_4 = y_8 = y_{10} = -1.2, & y_7 = y_{11} = -0.4, & y_9 = -0.8
\end{array}
$$

(3) Obtain a series of the form

$$b_1 \sin x + b_2 \sin 2x + b_3 \sin 3x$$

which gives an approximation to the function $y(x)$ which takes the same values as 10 sin $(x/4)$ at the points $x = 0(\pi/6)2\pi$.

(4) Find the values of a_0, a_1, a_2, b_1, b_2 such that the series $\frac{1}{2}a_0 + a_1 \cos x + a_2 \cos 2x + b_1 \sin x + b_2 \sin 2x$ takes the values given in the following table.

x	0°	30°	60°	90°	120°	150°
y	17	17	2	-8	2	17

x	180°	210°	240°	270°	300°	330°
y	17	7	2	2	2	7

(5) A function equals 1 in the range $0 \leqslant x \leqslant \pi$ and is zero in the range $\pi < x < 2\pi$. Compare the coefficients found in the Fourier series for this function by the method of integration and by the method of summation.

7
Numerical solution of equations

We consider first the solution of the equation $f(x) = 0$ by iterative methods. The accuracy obtainable in such methods is limited only by the round-off error in the last iteration. We need to examine the rate of convergence of various methods and estimate the amount of work involved. It saves time to begin by tabulating $f(x)$ at equal intervals of x in order to locate the roots approximately.

(a) The method of bisection

We find two points $x = x_1$, $x = x_2$ such that $f(x_1)$ and $f(x_2)$ have opposite signs. Then put $x_3 = \frac{1}{2}(x_1 + x_2)$ and find the sign of $f(x_3)$. If this is the same as the sign of $f(x_1)$, repeat the process by taking $x_4 = \frac{1}{2}(x_2 + x_3)$. Otherwise take $x_4 = \frac{1}{2}(x_1 + x_3)$. Assuming that $f(x)$ is continuous, we shall in this way obtain a sequence of points converging to a root of the equation $f(x) = 0$. This is simple but obviously slow.

(b) The method of false position

Again we begin with two points (x_1, y_1), (x_2, y_2) on the curve $y = f(x)$ such that y_1 and y_2 have opposite signs. The equation of the chord joining the two points is

$$(y - y_1)/(y_2 - y_1) = (x - x_1)/(x_2 - x_1)$$

so that this line cuts the x-axis at $x = x_3$ where

$$x_3 = (x_1 y_2 - x_2 y_1)/(y_2 - y_1) = x_2 - y_2(x_1 - x_2)/(y_1 - y_2)$$

In effect this amounts to inverse linear interpolation. We now find the value of $y_3 = f(x_3)$ and repeat the process using the points (x_2, y_2), (x_3, y_3) if y_2 and y_3 have opposite signs, or using (x_1, y_1), (x_3, y_3) if y_1 and y_3 have opposite signs. After a few iterations the slope of the chord changes very little, and it saves time to assign it a constant value.

To examine the rate of convergence we put $x_1 - \alpha = e_1$, $x_2 - \alpha = e_2$, $x_3 - \alpha = e_3$, where $x = \alpha$ is the exact root. Then from the equation of the chord we have

$$e_3 = (e_1 y_2 - e_2 y_1)/(y_2 - y_1)$$

If $f(x)$ can be expanded in powers of $(x - \alpha)$ we have

$$y_1 = f(\alpha) + (x_1 - \alpha)f'(\alpha) + \tfrac{1}{2}(x_1 - \alpha)^2 f''(\alpha) + \cdots$$

and similarly for y_2. Since $f(\alpha) = 0$ this gives approximately

$$e_3 = \tfrac{1}{2} e_1 e_2 f''(\alpha)/f'(\alpha)$$

We deduce that after n iterations

$$e_{n+2} = \tfrac{1}{2} e_n e_{n+1} f''(\alpha)/f'(\alpha)$$

This will fail if $f'(\alpha) = 0$, which will occur when the root is double. Apart from this case, the error e_3 will be proportional to the error e_2. For rapid convergence we need $f''(\alpha)/f'(\alpha)$ to be small.

If the curve $y = f(x)$ is known to be concave (or convex) between the points $P_1(x_1, y_1)$ and $P_2(x_2, y_2)$, we can employ a simpler alternative. If $P_1 P_2$ cuts the x-axis at $x = x_3$, we then find $y_3 = f(x_3)$, and use the chord joining the points (x_1, y_1), (x_3, y_3) to find the point $x = x_4$. We proceed in this way, keeping one end of the chord anchored at P_1. This gives an iteration defined by

$$x_{n+1} = x_n - y_n(x_n - x_1)/(y_n - y_1)$$

from which $e_{n+1} = \tfrac{1}{2} e_1 e_n f''(\alpha)/f'(\alpha)$ approximately.

For each of these methods we have approximately

$$f(x_n) = (x_n - \alpha)f'(\alpha) = (x_n - \alpha)(y_n - y_1)/(x_n - x_1)$$

so that

$$x_n - \alpha = f(x_n)(x_n - x_1)/(y_n - y_1)$$

This enables us to decide when to terminate the iteration. If we require the error in the result to be less than some quantity t, known as the tolerance, we continue the iteration until $|f(x_n)(x_n - x_1)/(y_n - y_1)|$ is less than t. Alternatively we can stop when $|f(x_{n+1})| \geqslant |f(x_n)|$.

(c) The secant method

The chord joining two points P_1 and P_2 on the curve $y = f(x)$ is used as in the method of false position to determine a point P_3 on the curve. We then use the points P_2 and P_3 to find P_4, and then use P_3 and P_4 to find P_5. We do not concern ourselves whether at any stage the two points we use are on opposite sides of the x-axis, so that there is a risk that the method may diverge. When the method is convergent, the advantage gained is that less work is entailed.

(*d*) The Newton–Raphson method

This is often used to improve a result obtained by one of the previous methods considered. In general, this method doubles the number of correct significant figures at each iteration. If $f(x)$ is differentiable in an interval containing the root $x = \alpha$ and an approximation $x = x_1$ to this root, we have by the mean value theorem

$$f(x_1) - f(\alpha) = (x_1 - \alpha)f'(\xi)$$

where ξ is some point between α and x_1. Since $f(\alpha) = 0$ this gives

$$\alpha = x_1 - f(x_1)/f'(\xi)$$

Since we do not know the value of ξ, we use $f'(x_1)$ as an approximation for $f'(\xi)$ and put

$$x_2 = x_1 - f(x_1)/f'(x_1)$$

This suggests the iteration

$$x_{n+1} = x_n - f(x_n)/f'(x_n)$$

In effect this is linear interpolation along the tangent to the curve $y = f(x)$ at the point (x_n, y_n). Since

$$f(\alpha) = f(x_n) + (\alpha - x_n)f'(x_n) + \tfrac{1}{2}(\alpha - x_n)^2 f''(x_n) + \cdots$$

we have

$$
\begin{aligned}
(x_{n+1} - \alpha) &= (x_n - \alpha) - f(x_n)/f'(x_n) \\
&= \tfrac{1}{2}(\alpha - x_n)^2 f''(x_n)/f'(x_n) + \cdots \\
e_{n+1} &= \tfrac{1}{2}e_n^2 f''(x_n)/f'(x_n) + \cdots \\
&= \tfrac{1}{2}e_n^2 f''(\alpha)/f'(\alpha) \quad \text{approximately.}
\end{aligned}
$$

i.e.,

Since e_{n+1} is proportional to the square of e_n, this is called a second-order method. The rate of convergence is higher than that of the methods above, but there is the disadvantage that the derivative $f'(x)$ has to be evaluated at several points. We can if we choose assign $f'(x)$ a suitable constant value, though this will reduce the rate of convergence. The difference $|x_{n+1} - x_n|$ diminishes rapidly as n increases and it is safe to assume that $|x_{n+1} - \alpha|$ is less than the tolerance t if $|f(x_n)/f'(x_n)| < t$. We terminate the iteration as soon as this inequality holds, or else as soon as $|f(x_{n+1})|$ is greater than $|f(x_n)|$.

An important application of this method is the evaluation of square roots and cube roots. Since it would not be economic to store tables of square roots and cube roots in a computer, it is essential to have ready subroutines for solving the equations $x^2 = a$, $x^3 = b$. If we apply the Newton–Raphson method to the equation

$$f(x) = x^2 - a = 0$$

we obtain

$$x_{n+1} = x_n - (x_n^2 - a)/2x_n = \tfrac{1}{2}(x_n + a/x_n)$$

If $x^2 = 17$ and $x_1 = 4$, this gives $x_2 = 4{\cdot}125$, $x_3 = 4{\cdot}1231 \ldots$, which is correct to four decimal places. Alternatively, we can use the equivalent form

$$f(x) = 1/x^2 - 1/a = 0$$

which gives

$$x_{n+1} = (3ax_n - x_n^3)/2a$$

This has the advantage that the divisor remains constant, though the rate of convergence is slightly slower. By a combination of these two iterations we can devise a third-order method, i.e., one in which $(x_{n+1} - \sqrt{a})$ is proportional to $(x_n - \sqrt{a})^3$. In practice, the gain in the rate of convergence is not worth the increased complexity of the iteration formula.

For the calculation of cube roots, we can use an iteration obtained by applying the Newton–Raphson method either to

(i) $f(x) = x^3 - a = 0$ or (ii) $f(x) = 1/x^3 - 1/a = 0$

The first gives the relation

$$x_{n+1} = \tfrac{1}{3}(2x_n + a/x_n^2)$$

and the second gives

$$x_{n+1} = \tfrac{1}{3}(4x_n - x_n^4/a)$$

To obtain an iteration for the calculation of a reciprocal, we apply the Newton–Raphson method to the equation

$$f(x) = 1/x - a = 0$$

We obtain

$$x_{n+1} = x_n - f(x_n)/f'(x_n) = 2x_n - ax_n^2$$

We can write this in the form

$$x_{n+1} - 1/a = -a(x_n - 1/a)^2$$

showing that, as expected, this is a second-order method. The rate of convergence is clearly dependent on the magnitude of the number a. If a is large, we can express it as the product of a power of 10 and a number b between 0 and 1. The reciprocal of b can be found by a few iterations and from it we can deduce the reciprocal of a. If the calculation is to be made in the binary scale, we express a as the product of a power of 2 and a number lying between 0 and 1.

(e) The method of successive substitutions

If the equation to be solved can be written in the form $x = F(x)$, it is possible that the iteration $x_{n+1} = F(x_n)$ will converge to a root $x = \alpha$. If $F(x)$ is differentiable, there is a number ξ between x_n and α such that

$$x_{n+1} - \alpha = F(x_n) - F(\alpha) = (x_n - \alpha)F'(\xi)$$

For convergence it is necessary that $|F'(\xi)|$ should be less than 1, and except for special cases in which $F'(\alpha) = 0$ this is a first-order method. The equation $x^3 + x^2 = 1$ clearly has a root between 0 and 1. If we put the equation into the form $x = (1 - x^2)/x^2$, the derivative of the right-hand side is $-2/x^3$. Since $|-2/x^3| > 1$ for $0 < x < 1$, the method of successive substitutions will fail. If instead we put the equation in the form $x = 1/\sqrt{(x + 1)}$, the derivative of the right-hand side is $-\frac{1}{2}(x + 1)^{-\frac{3}{2}}$, which has for positive x an absolute value less than 1. Thus the method will succeed with this form.

In cases where $|F'(\alpha)|$ is less than but nearly equal to unity, the convergence will be slow. We can accelerate the convergence by using Aitken's δ^2 process. Since $(x_{n+1} - \alpha)$ is approximately equal to $(x_n - \alpha)F'(\alpha)$, we have approximately

$$(x_{n+1} - \alpha)/(x_n - \alpha) = (x_n - \alpha)/(x_{n-1} - \alpha)$$

If we solve this relation for α, we find

$$\alpha = x_{n+1} - (x_{n+1} - x_n)^2/(x_{n+1} - 2x_n + x_{n-1})$$
$$= x_{n+1} - (\Delta x_n)^2/\delta^2 x_n$$

In this way, from three successive approximations x_{n-1}, x_n, x_{n+1} we can derive a much better approximation to the root α.

Matrix notation

For the proper understanding of the methods of solution of systems of linear equations, a knowledge of matrix notation is essential. An $(m \times n)$ matrix is a rectangular array of numbers with m rows and n columns. If $m = n$ it is a square matrix. The element of a matrix $\mathbf{A}$ in row i and column j is denoted by a_{ij}, so that we have

$$\mathbf{A} = \begin{bmatrix} a_{11} & a_{12} & \cdots & a_{1n} \\ a_{21} & a_{22} & \cdots & a_{2n} \\ \vdots & \vdots & \vdots & \vdots \\ a_{m1} & a_{m2} & \cdots & a_{mn} \end{bmatrix}$$

Two $(m \times n)$ matrices $\mathbf{A}$ and $\mathbf{B}$ are equal only if each element a_{ij} is equal to the corresponding element b_{ij}. The sum $\mathbf{A} + \mathbf{B}$ of two $(m \times n)$ matrices $\mathbf{A}$ and $\mathbf{B}$ is such that the element in its row i and its column j is $(a_{ij} + b_{ij})$. It follows that when a matrix is multiplied by a constant k, each element in the matrix is multiplied by k. The product $\mathbf{AB}$ of a $(p \times q)$ matrix $\mathbf{A}$ and a $(q \times r)$ matrix $\mathbf{B}$ is defined to be the $(p \times r)$ matrix in which the element in row i and column j is given by

$$a_{i1}b_{1j} + a_{i2}b_{2j} + a_{i3}b_{3j} + \cdots + a_{iq}b_{qj}$$

We calculate this element by multiplying each element in row i of $\mathbf{A}$ by the corresponding element in column j of $\mathbf{B}$, and then finding the sum of these products. Note that the product $\mathbf{AB}$ of a $(p \times q)$ matrix $\mathbf{A}$ and a $(q \times p)$ matrix $\mathbf{B}$ is not in general equal to the product $\mathbf{BA}$. Except when its determinant is zero, a square matrix $\mathbf{A}$ possesses an inverse $\mathbf{A}^{-1}$, such that $\mathbf{AA}^{-1} = \mathbf{A}^{-1}\mathbf{A} = \mathbf{I}$, where $\mathbf{I}$ is the unit matrix of the same order as $\mathbf{A}$. On the principal diagonal a unit matrix has unit elements, while all its other elements are zero. The eigenvalues (or latent roots) of a square matrix $\mathbf{A}$ are the values of λ for which the determinant of the matrix $(\mathbf{A} - \lambda\mathbf{I})$ is zero.

The following examples should be worked.

(1) $\quad 3\begin{bmatrix} 4 & 2 \\ 2 & -1 \end{bmatrix} - 2\begin{bmatrix} 5 & 2 \\ 1 & -3 \end{bmatrix} = \begin{bmatrix} 2 & 2 \\ 4 & 3 \end{bmatrix}$

(2) $\quad \begin{bmatrix} 2 & 3 & 1 \\ 3 & 4 & 2 \end{bmatrix} \begin{bmatrix} 3 & 1 & 3 \\ -2 & 0 & -1 \\ 1 & -2 & 2 \end{bmatrix} = \begin{bmatrix} 1 & 0 & 5 \\ 3 & -1 & 9 \end{bmatrix}$

(3) $\quad \begin{bmatrix} 1 & -3 & 5 \\ 7 & -7 & 11 \\ 3 & -6 & 10 \end{bmatrix} \begin{bmatrix} 1 \\ 5 \\ 3 \end{bmatrix} = \begin{bmatrix} 1 \\ 5 \\ 3 \end{bmatrix}$

(4) $\quad \begin{bmatrix} 1 \\ 2 \\ 3 \end{bmatrix} \begin{bmatrix} 1 & 2 & 3 \end{bmatrix} = \begin{bmatrix} 1 & 2 & 3 \\ 2 & 4 & 6 \\ 3 & 6 & 9 \end{bmatrix}$

(5) $\quad \begin{bmatrix} 1 & -1 & 1 \\ 2 & -1 & 0 \\ 1 & 0 & 0 \end{bmatrix} \begin{bmatrix} 0 & 0 & 1 \\ 0 & -1 & 2 \\ 1 & -1 & 1 \end{bmatrix} = \begin{bmatrix} 1 & 0 & 0 \\ 0 & 1 & 0 \\ 0 & 0 & 1 \end{bmatrix}$

(6) $\quad \begin{bmatrix} 1 & 2 \end{bmatrix} \begin{bmatrix} 3 & -4 \\ 5 & -6 \end{bmatrix} \begin{bmatrix} 3 \\ 2 \end{bmatrix} = 7$

In two dimensions we can represent a linear transformation by the equation $\mathbf{Ax} = \mathbf{y}$, where $\mathbf{A}$ is a (2×2) matrix. This transforms the vector $\mathbf{x}$ into the vector $\mathbf{y}$. If $\mathbf{A}$ is the matrix $\begin{bmatrix} -3 & 3 \\ -10 & 8 \end{bmatrix}$, then the vector $\begin{bmatrix} 1 \\ 1 \end{bmatrix}$ becomes the vector $\begin{bmatrix} 0 \\ -2 \end{bmatrix}$, and the vector $\begin{bmatrix} 4 \\ 5 \end{bmatrix}$ becomes the vector $\begin{bmatrix} 3 \\ 0 \end{bmatrix}$. This transformation is given by the equations

$$-3x_1 + 3x_2 = y_1, \qquad -10x_1 + 8x_2 = y_2$$

By solving for x_1 and x_2, we obtain the inverse transformation given by the equations

$$8y_1 - 3y_2 = 6x_1, \qquad 10y_1 - 3y_2 = 6x_2$$

which can be written in the form

$$\mathbf{A}^{-1}\mathbf{y} = \begin{bmatrix} \frac{4}{3} & -\frac{1}{2} \\ \frac{5}{3} & -\frac{1}{2} \end{bmatrix} \mathbf{y} = \mathbf{x}$$

Now two independent vectors can be found that are unchanged in direction by the transformation $\mathbf{Ax} = \mathbf{y}$. We have

$$\begin{bmatrix} -3 & 3 \\ -10 & 8 \end{bmatrix}\begin{bmatrix} 3 \\ 5 \end{bmatrix} = 2\begin{bmatrix} 3 \\ 5 \end{bmatrix}, \quad \text{and} \quad \begin{bmatrix} -3 & 3 \\ -10 & 8 \end{bmatrix}\begin{bmatrix} 1 \\ 2 \end{bmatrix} = 3\begin{bmatrix} 1 \\ 2 \end{bmatrix}$$

We say that the matrix $\mathbf{A}$ has eigenvalues 2 and 3, with eigenvectors $\begin{bmatrix} 3 \\ 5 \end{bmatrix}$ and $\begin{bmatrix} 1 \\ 2 \end{bmatrix}$ respectively. These eigenvalues could have been found by solving for λ the equation

$$|\mathbf{A} - \lambda\mathbf{I}| = \begin{vmatrix} -3 - \lambda & 3 \\ -10 & 8 - \lambda \end{vmatrix} = 0$$

An example of a linear transformation in three dimensions is given by

$$\mathbf{Bx} = \begin{bmatrix} 2 & -2 & 3 \\ 1 & 1 & 1 \\ 1 & 3 & -1 \end{bmatrix} \mathbf{x} = \mathbf{y}$$

From the equation

$$\begin{vmatrix} 2 - \lambda & -2 & 3 \\ 1 & 1 - \lambda & 1 \\ 1 & 3 & -1 - \lambda \end{vmatrix} = 0$$

we find that the eigenvalues of $\mathbf{B}$ are 3, -2, 1. Later we shall meet the matrix

$$
\mathbf{C} = \begin{bmatrix} 0 & 0 & \frac{1}{4} & \frac{1}{4} \\ 0 & 0 & \frac{1}{4} & \frac{1}{4} \\ \frac{1}{4} & \frac{1}{4} & 0 & 0 \\ \frac{1}{4} & \frac{1}{4} & 0 & 0 \end{bmatrix}
$$

and we shall be interested in the sequence $\mathbf{Cu}$, $\mathbf{C}^2\mathbf{u}$, $\mathbf{C}^3\mathbf{u}$, ..., where $\mathbf{u}$ is a vector representing an error. The matrix $\mathbf{C}$ has eigenvalues 0, 0, $\frac{1}{2}$, $-\frac{1}{2}$, with eigenvectors

$$
\mathbf{u}_1 = \begin{bmatrix} 1 \\ -1 \\ 0 \\ 0 \end{bmatrix},\ \mathbf{u}_2 = \begin{bmatrix} 0 \\ 0 \\ 1 \\ -1 \end{bmatrix},\ \mathbf{u}_3 = \begin{bmatrix} 1 \\ 1 \\ 1 \\ 1 \end{bmatrix},\ \mathbf{u}_4 = \begin{bmatrix} 1 \\ 1 \\ -1 \\ -1 \end{bmatrix}
$$

We can express any vector $\mathbf{u}$ as a linear combination of multiples of these eigenvectors, say

$$
\mathbf{u} = a_1\mathbf{u}_1 + a_2\mathbf{u}_2 + a_3\mathbf{u}_3 + a_4\mathbf{u}_4
$$

Then the elements of the vector $\mathbf{C}^n\mathbf{u}$ will tend to zero as n increases, for

$$
\mathbf{Cu} = \tfrac{1}{2}a_3\mathbf{u}_3 - \tfrac{1}{2}a_4\mathbf{u}_4, \quad \text{and} \quad \mathbf{C}^n\mathbf{u} = (\tfrac{1}{2})^n a_3\mathbf{u}_3 + (-\tfrac{1}{2})^n a_4\mathbf{u}_4
$$

Sets of linear equations

We are faced with the problem of solving a set of n linear equations in n unknowns, which can be set out in the form

$$
\begin{aligned}
a_{11}x_1 + a_{12}x_2 + \cdots a_{1n}x_n &= k_1 \\
a_{21}x_1 + a_{22}x_2 + \cdots a_{2n}x_n &= k_2 \\
&\ \ \vdots \\
a_{n1}x_1 + a_{n2}x_2 + \cdots a_{nn}x_n &= k_n
\end{aligned}
$$

There are two distinct modes of attack, the direct methods which give a result in a fixed number of steps, and the iterative methods in which a process is repeated until the accuracy required is achieved. We have to bear in mind the necessity for checking the calculations at each stage, and also the risk of serious inaccuracy caused by round-off errors. The coefficients in the equations will be decimal numbers, each rounded-off say to t decimal places, and we are trying to obtain values of the unknowns x_1, x_2, ..., x_n to the same number of decimal places. Thus we cannot hope that our solution will satisfy the equations exactly.

We consider first the direct method of pivotal condensation or Gaussian elimination. We begin by multiplying each row by a suitable factor, so that the largest coefficient in the row lies between $0 \cdot 1$ and 1, a process known as scaling. If the array of coefficients is symmetric, we maintain the symmetry by treating the columns in the same way. Next by interchange of rows and columns we bring the largest coefficient to the top left-hand corner of the array. The coefficients are then set out (without the variables $x_1, x_2, \ldots, x_n$) with a column containing the constants $k_1, k_2, \ldots, k_n$. If the calculation is by hand-machine, we include an extra column $s_1, s_2, \ldots, s_n$, in which s_r gives the sum of the elements in the rth row. We then find the sum of each column.

For $n = 4$ we have the array

$$
\begin{array}{cccccc}
a_{11} & a_{12} & a_{13} & a_{14} & k_1 & s_1 \\
a_{21} & a_{22} & a_{23} & a_{24} & k_2 & s_2 \\
a_{31} & a_{32} & a_{33} & a_{34} & k_3 & s_3 \\
a_{41} & a_{42} & a_{43} & a_{44} & k_4 & s_4 \\
S_1 & S_2 & S_3 & S_4 & S_5 & S
\end{array}
$$

We now check that $S_1 + S_2 + S_3 + S_4 + S_5 = S$.

In the first stage we aim at eliminating the first element from all rows except the first. The element a_{11} is called the first pivot, and we use it to calculate the multipliers m_{21}, m_{31}, m_{41} defined by

$$
m_{r1} = -a_{r1}/a_{11}, \quad r = 2, 3, 4
$$

We now add the elements of row 1 multiplied by m_{21} to the elements of row 2, so that a_{21} is replaced by zero. In the same way, a_{31} and a_{41} are replaced by zero. We now have an array

$$
\begin{array}{cccccc}
a_{11} & a_{12} & a_{13} & a_{14} & k_1 & s_1 \\
0 & b_{22} & b_{23} & b_{24} & k_5 & s_5 \\
0 & b_{32} & b_{33} & b_{34} & k_6 & s_6 \\
0 & b_{42} & b_{43} & b_{44} & k_7 & s_7
\end{array}
$$

The value of each of the row sums s_5, s_6, s_7 is obtained by two routes. For example, we have

$$
s_5 = s_2 + m_{21}s_1, \quad \text{and} \quad s_5 = b_{22} + b_{23} + b_{24} + k_5
$$

Apart from a possible disagreement in the final digit, the two answers should be the same, giving a very useful check. When the final digits differ, the value found from the row sum is to be taken.

From the three numbers b_{22}, b_{32}, b_{42} we select the largest in magnitude for our second pivot. Assuming that this is b_{22}, we then form the multipliers m_{32} and m_{42} where

$$
m_{r2} = -b_{r2}/b_{22}, \quad r = 3, 4
$$

We now add the elements of row 2 multiplied by m_{32} to row 3, and we add the elements of row 2 multiplied by m_{42} to row 4, to obtain zeros in place of b_{32} and b_{42}. The new array can be written

$$
\begin{array}{cccccc}
a_{11} & a_{12} & a_{13} & a_{14} & k_1 & s_1 \\
0 & b_{22} & b_{23} & b_{24} & k_5 & s_5 \\
0 & 0 & c_{33} & c_{34} & k_8 & s_8 \\
0 & 0 & c_{43} & c_{44} & k_9 & s_9
\end{array}
$$

Our third pivot is then whichever of c_{33} and c_{43} is larger in magnitude. If it is c_{33}, we put

$$
m_{43} = -c_{43}/c_{33}
$$

and add the elements of row 3 multiplied by m_{43} to row 4. The final array is then

$$
\begin{array}{cccccc}
a_{11} & a_{12} & a_{13} & a_{14} & k_1 & s_1 \\
0 & b_{22} & b_{23} & b_{24} & k_5 & s_5 \\
0 & 0 & c_{33} & c_{34} & k_8 & s_8 \\
0 & 0 & 0 & d_{44} & k_{10} & s_{10}
\end{array}
$$

It is important to notice that by our choice of pivots we have made sure that each multiplier lies between -1 and 1. This helps to prevent any element becoming greater than 1 in magnitude, and it also reduces the error arising from round-off. Our system of equations has now been reduced to a triangular system:

$$
\begin{aligned}
a_{11}x_1 + a_{12}x_2 + a_{13}x_3 + a_{14}x_4 &= k_1 \\
b_{22}x_2 + b_{23}x_3 + b_{24}x_4 &= k_5 \\
c_{33}x_3 + c_{34}x_4 &= k_8 \\
d_{44}x_4 &= k_{10}
\end{aligned}
$$

This system is solved immediately by the method of back-substitution:

$$
\begin{aligned}
x_4 &= k_{10}/d_{44} \\
x_3 &= (k_8 - c_{34}x_4)/c_{33} \\
x_2 &= (k_5 - b_{23}x_3 - b_{24}x_4)/b_{22} \\
x_1 &= (k_1 - a_{12}x_2 - a_{13}x_3 - a_{14}x_4)/a_{11}
\end{aligned}
$$

We round off the values of x_1, x_2, x_3, and x_4 to t decimal places. Now if we add the original four equations together we obtain the equation

$$
S_1 x_1 + S_2 x_2 + S_3 x_3 + S_4 x_4 = S_5
$$

We substitute the values found for x_1, x_2, x_3, and x_4 in the left-hand side of this equation. The result should agree with S_5 or differ only in the last digit.

If n is not greater than 10, it is usual to keep one guard figure, i.e., to use $(t + 1)$ decimal places throughout when using a desk machine. If n is between 10 and 100, two guard figures are necessary. This will reduce the error caused by round-off, which arises in three ways. First, the values of the multipliers are rounded-off values so that m_{21} is not exactly equal to $-a_{21}/a_{11}$ and so forth. Second, we multiply a_{12}, a_{13}, and a_{14} by m_{21} and round off these products before adding to the corresponding element in row 2. Third, round-off error occurs in the back-substitution.

In order to reduce this error, several compact methods have been devised. We have space for only one of these, known as Crout's method. With $n = 4$, and the same system of equations as above, we begin by dividing the elements in row 1 by a_{11}, to give a new row 1:

$$1 \qquad a'_{12} \qquad a'_{13} \qquad a'_{14} \qquad k'_1$$

We then proceed as follows:

$$a'_{22} = a_{22} - a_{21}a'_{12}, \qquad a'_{23} = (a_{23} - a_{21}a'_{13})/a'_{22}$$
$$a'_{32} = a_{32} - a_{31}a'_{12}, \qquad a'_{24} = (a_{24} - a_{21}a'_{14})/a'_{22}$$
$$a'_{42} = a_{42} - a_{41}a'_{12}, \qquad k'_2 = (k_2 - a_{21}k'_1)/a'_{22}$$

$$a'_{33} = a_{33} - a_{31}a'_{13} - a'_{32}a'_{23}$$
$$a'_{34} = (a_{34} - a_{31}a'_{14} - a'_{32}a'_{24})/a'_{33}$$
$$k'_3 = (k_3 - a_{31}k'_1 - a'_{32}k'_2)/a'_{33}$$
$$a'_{43} = a_{43} - a_{41}a'_{13} - a'_{42}a'_{23}$$
$$a'_{44} = a_{44} - a_{41}a'_{14} - a'_{42}a'_{24} - a'_{43}a'_{34}$$
$$k'_4 = (k_4 - a_{41}k'_1 - a'_{42}k'_2 - a'_{43}k'_3)/a'_{44}$$

The final array can now be set out as follows:

$$
\begin{array}{ccccc}
1 & a'_{12} & a'_{13} & a'_{14} & k'_1 \\
 & 1 & a'_{23} & a'_{24} & k'_2 \\
 & & 1 & a'_{34} & k'_3 \\
 & & & 1 & k'_4
\end{array}
$$

$$
\begin{array}{cccc}
a_{11} & & & \\
a_{21} & a'_{22} & & \\
a_{31} & a'_{32} & a'_{33} & \\
a_{41} & a'_{42} & a'_{43} & a'_{44}
\end{array}
$$

The upper triangle of elements corresponds to the triangular system of equations

$$x_1 + a'_{12}x_2 + a'_{13}x_3 + a'_{14}x_4 = k'_1$$
$$x_2 + a'_{23}x_3 + a'_{24}x_4 = k'_2$$
$$x_3 + a'_{34}x_4 = k'_3$$
$$x_4 = k'_4$$

We can now solve this system by the method of back-substitution. The obvious advantage of this compact method is that we change each element in the array once only, and in doing so we minimize the round-off error by accumulating products. The sum column for the current checks has been omitted above, but it should be included if the calculation is made on a desk machine.

Using the calculated values of the unknowns rounded-off to the required number of decimal places, we put

$$r_1 = k_1 - (a_{11}x_1 + a_{12}x_2 + a_{13}x_3 + a_{14}x_4)$$
$$r_2 = k_2 - (a_{21}x_1 + a_{22}x_2 + a_{23}x_3 + a_{24}x_4)$$
$$r_3 = k_3 - (a_{31}x_1 + a_{32}x_2 + a_{33}x_3 + a_{34}x_4)$$
$$r_4 = k_4 - (a_{41}x_1 + a_{42}x_2 + a_{43}x_3 + a_{44}x_4)$$

The numbers r_1, r_2, r_3, r_4 are known as the residuals. If we have made no blunders, their values will be small. It is a mistake however to assume that if the residuals are small, then the solution found is close to the true solution. Consider for example the two simultaneous equations

$$0{\cdot}100x + 0{\cdot}165y = 0{\cdot}166$$
$$0{\cdot}067x + 0{\cdot}110y = 0{\cdot}111$$

with the exact solution $x = 1$, $y = 0{\cdot}4$.

If we put $x = 0{\cdot}001$ and $y = 1{\cdot}000$, we find

$$0{\cdot}100x + 0{\cdot}165y - 0{\cdot}166 = 0$$
$$0{\cdot}067x + 0{\cdot}110y - 0{\cdot}111 = -0{\cdot}00033$$

The residuals are small, but the solution is absurdly wrong. Geometrically we can interpret this as the problem of finding the point of intersection of two straight lines which are almost parallel to one another. It is plain that a very small change in any coefficient will produce a large change in the solution. A problem which has this characteristic is said to be ill-conditioned. This difficulty arises in the solution of a system of linear equations when the value of the determinant of the coefficients of the unknowns is very small. In this case, small changes in the coefficients can have a great effect on the solution, and the same is true of the errors due to round-off. This ill-condition shows itself in the decrease in the magnitude of the pivots, and a loss of significant figures in the pivot will lead to a corresponding loss in the solution.

If we replace the constants k_1, k_2, k_3, and k_4 by the residuals r_1, r_2, r_3, and r_4 respectively and solve the system for the second time, we shall obtain corrections to the values already obtained for the unknowns. For if $\mathbf{x}$ is the calculated solution, then in matrix notation $\mathbf{r} = \mathbf{k} - \mathbf{A}\mathbf{x}$ so that if $\mathbf{A}\mathbf{y} = \mathbf{r}$ we have $\mathbf{A}(\mathbf{x} + \mathbf{y}) = \mathbf{k}$. The amount of extra work entailed will not be large.

We now solve the system below by our two methods.

$$0.41x_1 + 0.10x_2 + 0.11x_3 + 0.12x_4 = 0.13 \quad (s_1 = 0.87)$$
$$0.10x_1 + 0.31x_2 + 0.09x_3 + 0.09x_4 = 0.14 \quad (s_2 = 0.73)$$
$$0.21x_1 + 0.10x_2 + 0.30x_3 + 0.09x_4 = 0.16 \quad (s_3 = 0.86)$$
$$0.09x_1 + 0.21x_2 + 0.11x_3 + 0.21x_4 = 0.16 \quad (s_4 = 0.78)$$

(*a*) Pivotal condensation

We keep one guard figure

	0·410	0·100	0·110	0·120	0·130	0·870
$m_{21} = -0.244$		0·286	0·063	0·061	0·108	0·518
$m_{31} = -0.512$		0·049	0·244	0·029	0·093	0·415
$m_{41} = -0.220$		0·188	0·086	0·184	0·131	0·589
$m_{32} = -0.171$			0·233	0·019	0·075	0·327
$m_{42} = -0.657$			0·045	0·144	0·060	0·249
$m_{43} = -0.193$				0·140	0·046	0·186

The numbers underlined are the pivots. We now have the triangular system

$$0.410x_1 + 0.100x_2 + 0.110x_3 + 0.120x_4 = 0.130$$
$$0.286x_2 + 0.063x_3 + 0.061x_4 = 0.108$$
$$0.233x_3 + 0.019x_4 = 0.075$$
$$0.140x_4 = 0.046$$

By back-substitution we find

$$x_4 = 0.329, \quad x_3 = 0.296, \quad x_2 = 0.242, \quad x_1 = 0.081$$

The values rounded-off to two decimal places are

$$x_1 = 0.08, \quad x_2 = 0.24, \quad x_3 = 0.30, \quad x_4 = 0.33$$

If we add the original four equations together, we obtain the equation

$$0.81x_1 + 0.72x_2 + 0.61x_3 + 0.51x_4 = 0.59$$

We now substitute our values in the left-hand side and find the value 0·5889. The corresponding residuals are

$$r_1 = 0.0006, \quad r_2 = 0.0009, \quad r_3 = -0.0005, \quad r_4 = 0.0001$$

(*b*) Crout's method

The initial array is as before. The final array becomes

$$
\begin{array}{cccccc}
1 & 0\cdot244 & 0\cdot268 & 0\cdot293 & 0\cdot317 & 2\cdot122 \\
 & 1 & 0\cdot220 & 0\cdot213 & 0\cdot378 & 1\cdot811 \\
 & & 1 & 0\cdot077 & 0\cdot322 & 1\cdot399 \\
 & & & 1 & 0\cdot329 & 1\cdot329 \\
0\cdot41 & & & & & \\
0\cdot10 & 0\cdot286 & & & & \\
0\cdot21 & 0\cdot049 & 0\cdot233 & & & \\
0\cdot09 & 0\cdot188 & 0\cdot045 & 0\cdot140 & &
\end{array}
$$

This has been calculated using accumulation of products, with the final value of each element rounded-off to three decimal places. Thus

$$a'_{34} = \{0\cdot09 - (0\cdot21)(0\cdot293) - (0\cdot049)(0\cdot213)\}/(0\cdot233) = 0\cdot077$$

We have now the triangular system of equations

$$
\begin{aligned}
x_1 + 0\cdot244x_2 + 0\cdot268x_3 + 0\cdot293x_4 &= 0\cdot317 \\
x_2 + 0\cdot220x_3 + 0\cdot213x_4 &= 0\cdot378 \\
x_3 + 0\cdot077x_4 &= 0\cdot322 \\
x_4 &= 0\cdot329
\end{aligned}
$$

By back-substitution we obtain the solution

$$x_4 = 0\cdot329, \qquad x_3 = 0\cdot297, \qquad x_2 = 0\cdot243, \qquad x_1 = 0\cdot082$$

After rounding-off to two decimal places, we have the same result as before.

It must be pointed out that this solution will be unreliable if the coefficients in the original system of equations are not exact values, i.e., if they are rounded-off or if they come from experimental results. If the coefficient of x_4 in the fourth equation is changed from $0\cdot21$ to $0\cdot206$, it is easy to show that the solution gives x_4 equal to $0\cdot338$ instead of $0\cdot329$. This emphasizes the fact that when the given coefficients are subject to uncertainty, it is pointless and misleading to quote a solution to a spurious degree of accuracy.

We can compare the two methods used above in matrix notation. We write the system of equations in matrix form

$$\mathbf{Ax = k}$$

where the matrix $\mathbf{A}$ is

$$
\begin{bmatrix}
0\cdot41 & 0\cdot10 & 0\cdot11 & 0\cdot12 \\
0\cdot10 & 0\cdot31 & 0\cdot09 & 0\cdot09 \\
0\cdot21 & 0\cdot10 & 0\cdot30 & 0\cdot09 \\
0\cdot09 & 0\cdot21 & 0\cdot11 & 0\cdot21
\end{bmatrix}
$$

while $\mathbf{x}$ and $\mathbf{k}$ are the column vectors

$$(x_1, x_2, x_3, x_4)' \quad \text{and} \quad (0\cdot13, 0\cdot14, 0\cdot16, 0\cdot16)'$$

In the solution by pivotal condensation, we express the matrix $\mathbf{A}$ as the product of a lower triangular matrix $\mathbf{L}_1$ and an upper triangular matrix $\mathbf{U}_1$, with unit elements in the diagonal of $\mathbf{L}_1$:

$$\mathbf{L}_1 = \begin{bmatrix} 1 & & & \\ 0{\cdot}244 & 1 & & \\ 0{\cdot}512 & 0{\cdot}171 & 1 & \\ 0{\cdot}220 & 0{\cdot}657 & 0{\cdot}193 & 1 \end{bmatrix}$$

$$\mathbf{U}_1 = \begin{bmatrix} 0{\cdot}410 & 0{\cdot}100 & 0{\cdot}110 & 0{\cdot}120 \\ & 0{\cdot}286 & 0{\cdot}063 & 0{\cdot}061 \\ & & 0{\cdot}233 & 0{\cdot}019 \\ & & & 0{\cdot}140 \end{bmatrix}$$

In Crout's method, we express the matrix $\mathbf{A}$ as the product of a lower triangular matrix $\mathbf{L}_2$ and an upper triangular matrix $\mathbf{U}_2$, with unit elements on the diagonal of $\mathbf{U}_2$:

$$\mathbf{L}_2 = \begin{bmatrix} 0{\cdot}410 & & & \\ 0{\cdot}100 & 0{\cdot}286 & & \\ 0{\cdot}210 & 0{\cdot}049 & 0{\cdot}233 & \\ 0{\cdot}090 & 0{\cdot}188 & 0{\cdot}045 & 0{\cdot}140 \end{bmatrix}$$

$$\mathbf{U}_2 = \begin{bmatrix} 1 & 0{\cdot}244 & 0{\cdot}268 & 0{\cdot}293 \\ & 1 & 0{\cdot}220 & 0{\cdot}213 \\ & & 1 & 0{\cdot}077 \\ & & & 1 \end{bmatrix}$$

Notice that if we divide a column in $\mathbf{L}_2$ by the diagonal element in that column we obtain the corresponding column in $\mathbf{L}_1$, and that if we divide a row in $\mathbf{U}_1$ by the diagonal element in that row we obtain the corresponding row in $\mathbf{U}_2$.

Clearly, neither $\mathbf{L}_1\mathbf{U}_1$ nor $\mathbf{L}_2\mathbf{U}_2$ is exactly equal to $\mathbf{A}$. It is interesting to find the discrepancy in each case.

$$\mathbf{A} - \mathbf{L}_1\mathbf{U}_1 = 10^{-6} \times \begin{bmatrix} 0 & 0 & 0 & 0 \\ -40 & -400 & 160 & -280 \\ 80 & -126 & -219 & -876 \\ -200 & 98 & -560 & -144 \end{bmatrix}$$

$$\mathbf{A} - \mathbf{L}_2\mathbf{U}_2 = 10^{-6} \times \begin{bmatrix} 0 & -40 & 120 & -130 \\ 0 & -400 & 280 & -218 \\ 0 & -240 & -60 & 542 \\ 0 & 40 & -450 & 121 \end{bmatrix}$$

In both cases the two largest elements in the error matrix correspond to the elements a_{34} and a_{43} in $\mathbf{A}$.

Next we consider the following system, which is more typical than the previous example:

$$0{\cdot}5121x_1 + 0{\cdot}3241x_2 + 0{\cdot}3123x_3 + 0{\cdot}4216x_4 = 0{\cdot}5927$$
$$0{\cdot}4266x_1 + 0{\cdot}8312x_2 + 0{\cdot}2912x_3 + 0{\cdot}3120x_4 = 0{\cdot}6021$$
$$0{\cdot}1312x_1 + 0{\cdot}6141x_2 + 0{\cdot}3232x_3 + 0{\cdot}4122x_4 = 0{\cdot}3672$$
$$0{\cdot}3461x_1 + 0{\cdot}2003x_2 + 0{\cdot}4132x_3 + 0{\cdot}2631x_4 = 1{\cdot}1522$$

We solve this system by the method of pivotal condensation, the pivots being underlined. In the layout below, the multipliers are given in the first column and the row sums in the last column. We work to five decimal places.

	0·5121	0·3241	0·3123	0·4216	0·5927	2·1628
−0·83304	0·4266	0·8312	0·2912	0·3120	0·6021	2·4631
−0·25620	0·1312	0·6141	0·3232	0·4122	0·3672	1·8479
−0·67584	0·3461	0·2003	0·4132	0·2631	1·1522	2·3749
		0·56121	0·03104	−0·03921	0·10836	0·66140
−0·94629		0·53107	0·24319	0·30419	0·21535	1·29380
0·03339		−0·01874	0·20214	−0·02183	0·75163	0·91320
			0·21382	0·34129	0·11281	0·66792
−0·95024			0·20318	−0·02314	0·75525	0·93529
				−0·34745	0·64805	0·30060

By back substitution we find

$$x_4 = -1{\cdot}86516, \quad x_3 = 3{\cdot}50468, \quad x_2 = -0{\cdot}13107, \quad x_1 = 0{\cdot}63859$$

We round these values off to four decimal places, giving the solution

$$x_1 = 0{\cdot}6386, \quad x_2 = -0{\cdot}1311, \quad x_3 = 3{\cdot}5047, \quad x_4 = -1{\cdot}8652$$

The sum of the four equations is the equation

$$1{\cdot}4160x_1 + 1{\cdot}9697x_2 + 1{\cdot}3399x_3 + 1{\cdot}4089x_4 = 2{\cdot}7142$$

If we substitute our result in the left-hand side we obtain $2{\cdot}7141$.

EXAMPLES

(1) Solve the following system of equations by the pivotal condensation method and by Crout's method:

$$
\begin{aligned}
4x_1 + 2x_2 + x_3 + 3x_4 &= 2\cdot 8 \\
-2x_1 + 3x_2 - 1\cdot 5x_3 - 3\cdot 5x_4 &= -2\cdot 2 \\
3x_1 + 2\cdot 5x_2 + 3\cdot 5x_3 + 2\cdot 75x_4 &= 4\cdot 7 \\
2x_1 + 5x_2 - 1\cdot 75x_3 + 0\cdot 5x_4 &= 0
\end{aligned}
$$

(2) Solve the following system, working with three decimal places:

$$
\begin{aligned}
0\cdot 50x_1 - 0\cdot 11x_2 - 0\cdot 24x_3 + 0\cdot 16x_4 &= 0\cdot 197 \\
0\cdot 12x_1 + 0\cdot 48x_2 - 0\cdot 31x_3 + 0\cdot 14x_4 &= 0\cdot 230 \\
0\cdot 10x_1 + 0\cdot 31x_2 + 0\cdot 61x_3 - 0\cdot 18x_4 &= 0\cdot 405 \\
0\cdot 25x_1 + 0\cdot 12x_2 + 0\cdot 26x_3 + 0\cdot 54x_4 &= 0\cdot 476
\end{aligned}
$$

(3) Solve the system

$$
\begin{aligned}
0\cdot 6333x_1 - 0\cdot 3142x_2 - 0\cdot 3653x_3 &= 0\cdot 1028 \\
0\cdot 0604x_1 + 0\cdot 7400x_2 + 0\cdot 9659x_3 &= 1\cdot 2138 \\
-0\cdot 0812x_1 + 0\cdot 0405x_2 + 0\cdot 2814x_3 &= 0\cdot 1311
\end{aligned}
$$

(4) In the simultaneous equations

$$
0\cdot 4x + 0\cdot 6y = 0\cdot 3, \qquad 0\cdot 3x + 0\cdot 5y = 0\cdot 1
$$

the coefficients of x and y have been rounded-off. Show that no solution is reliable.

(5) Find the residuals when the values $x_1 = 1\cdot 082$, $x_2 = 0\cdot 864$, $x_3 = 1\cdot 035$, $x_4 = 0\cdot 979$ are substituted in the system of equations

$$
\begin{aligned}
3x_1 + 2x_2 + 2x_3 + 2x_4 &= 9 \\
x_1 + x_2 + x_3 - x_4 &= 2 \\
x_1 + x_2 + 4x_3 + 4x_4 &= 10 \\
0x_1 + 0x_2 + 3x_3 + 5x_4 &= 8
\end{aligned}
$$

Solve the system by pivotal condensation using exact arithmetic.

Iterative methods

In the numerical solution of partial differential equations it is often necessary to deal with a system of linear simultaneous equations in which many co-efficients are zero. In such a case, it may be possible to use an iterative method of solution. The outstanding advantage of such a method is that the rounding-off errors do not accumulate. It is only at the last step that any rounding-off error affects the final result. We begin by writing the system of equations in the form

$$x_1 = (k_1 - a_{12}x_2 - a_{13}x_3 - a_{14}x_4)/a_{11}$$
$$x_2 = (k_2 - a_{21}x_1 - a_{23}x_3 - a_{24}x_4)/a_{22}$$
$$x_3 = (k_3 - a_{31}x_1 - a_{32}x_2 - a_{34}x_4)/a_{33}$$
$$x_4 = (k_4 - a_{41}x_1 - a_{42}x_2 - a_{43}x_3)/a_{44}$$

Each equation is used to give a value for the variable on the left-hand side. For our initial approximations we can take

$$x_1 = k_1/a_{11}, \quad x_2 = k_2/a_{22}, \quad x_3 = k_3/a_{33}, \quad x_4 = k_4/a_{44}$$

especially when the coefficients $a_{11}, a_{22}, a_{33}, a_{44}$ are much larger in magnitude than the other coefficients. Two basic methods are known as Jacobi's method and the Gauss–Seidel method. In the first of these, we substitute our initial values in the right-hand side of the equations and hence obtain the next approximation. We repeat this process until two successive approximations agree to the degree of accuracy required (or until we recognize that the process is failing to converge). In the Gauss–Seidel method we use the equations one at a time. We begin by using the first equation to revise our approximation for x_1, and then we use this new value for x_1 immediately in the second equation to give a new value for x_2. Each new value is employed as soon as it is available, whereas in Jacobi's method the new values are all introduced at the same point, i.e., at the commencement of another cycle.

Both of these iterative methods will certainly converge if the coefficient of x_r in the rth row is greater in magnitude than the sum of the moduli of the other coefficients in that row. This condition is sufficient for convergence, but it is possible for the iteration to converge when this condition is not satisfied. From the practical point of view, convergence alone is not enough; what is required is rapid convergence.

Consider the system

$$4x_1 - x_3 - x_4 = 4$$
$$4x_2 - x_3 - x_4 = 4$$
$$x_1 + x_2 - 4x_3 = -4$$
$$x_1 + x_2 - 4x_4 = -4$$

If we take $x_1 = x_2 = x_3 = x_4 = 1$ for our initial approximation, the Jacobi method gives the sequence of sets of values

(a) $x_1 = x_2 = x_3 = x_4 = 1\frac{1}{2}$
(b) $x_1 = x_2 = x_3 = x_4 = 1\frac{3}{4}$
(c) $x_1 = x_2 = x_3 = x_4 = 1\frac{7}{8}$

After n iterations we reach the value $2 - (\frac{1}{2})^n$ for each x. The Gauss–Seidel method, with the same initial approximation gives

(a) $x_1 = 1\frac{1}{2}, x_2 = 1\frac{1}{2}, x_3 = 1\frac{3}{4}, x_4 = 1\frac{3}{4}$

(b) $x_1 = 1\frac{7}{8}, x_2 = 1\frac{7}{8}, x_3 = 1\frac{15}{16}, x_4 = 1\frac{15}{16}$

It is clear that the rate of convergence is twice as high in the second of the two methods as in the first. There are however cases in which the Jacobi method converges while the Gauss–Seidel method diverges.

The difference between these two methods can be expressed clearly in matrix notation. We split the matrix $\mathbf{A}$ of the coefficients into the diagonal matrix $\mathbf{D}$ and the triangular matrices $\mathbf{L}$ and $\mathbf{U}$ (with zeros on the diagonals). The system of equations $\mathbf{Ax} = \mathbf{k}$ is then written

$$(\mathbf{L} + \mathbf{D} + \mathbf{U})\mathbf{x} = \mathbf{k}$$

In the Jacobi method, two successive approximations $\mathbf{x}^{(i)}$ and $\mathbf{x}^{(i+1)}$ are linked by the relation

$$\mathbf{Dx}^{(i+1)} = \mathbf{k} - (\mathbf{L} + \mathbf{U})\mathbf{x}^{(i)}$$

so that the error in these approximations satisfies

$$(\mathbf{x}^{(i+1)} - \mathbf{x}) = -\mathbf{D}^{-1}(\mathbf{L} + \mathbf{U})(\mathbf{x}^{(i)} - \mathbf{x})$$

The matrix $-\mathbf{D}^{-1}(\mathbf{L} + \mathbf{U})$ is known as the iteration matrix for this method. In the example above we have

$$-\mathbf{D}^{-1}(\mathbf{L} + \mathbf{U}) = \begin{bmatrix} 0 & 0 & \frac{1}{4} & \frac{1}{4} \\ 0 & 0 & \frac{1}{4} & \frac{1}{4} \\ \frac{1}{4} & \frac{1}{4} & 0 & 0 \\ \frac{1}{4} & \frac{1}{4} & 0 & 0 \end{bmatrix}$$

The four eigenvalues of this matrix are $0, 0, \frac{1}{2}, -\frac{1}{2}$. Since none of these exceeds 1 in magnitude, the method will converge in this example.

In the Gauss–Seidel method, two successive approximations are linked by the relation

$$(\mathbf{L} + \mathbf{D})\mathbf{x}^{(i+1)} = \mathbf{k} - \mathbf{Ux}^{(i)}$$

so that

$$(\mathbf{x}^{(i+1)} - \mathbf{x}) = -(\mathbf{L} + \mathbf{D})^{-1}\mathbf{U}(\mathbf{x}^{(i)} - \mathbf{x})$$

The iteration matrix here is $-(\mathbf{L} + \mathbf{D})^{-1}\mathbf{U}$, and in the example above we have

$$-(\mathbf{L} + \mathbf{D})^{-1}\mathbf{U} = \begin{bmatrix} 0 & 0 & \frac{1}{4} & \frac{1}{4} \\ 0 & 0 & \frac{1}{4} & \frac{1}{4} \\ 0 & 0 & \frac{1}{8} & \frac{1}{8} \\ 0 & 0 & \frac{1}{8} & \frac{1}{8} \end{bmatrix}$$

Since the eigenvalues of this matrix are $0, 0, 0, \frac{1}{4}$, we can see why the rate of convergence with the Gauss–Seidel method in the example was faster than in the Jacobi method. The rate of convergence depends essentially on the magnitude of the largest eigenvalue of the iteration matrix. In a large system it is not uncommon for this to lie between 0·99 and 1, and much effort has been spent in recent years in devising ways of accelerating the convergence of iterative methods.

EXAMPLES

(1) Apply the Jacobi method and the Gauss–Seidel method to the system

$$
\begin{aligned}
x_1 \quad\quad\quad - \quad x_3 &= \quad 1 \\
x_1 - \quad x_2 \quad\quad\quad &= -1 \\
x_1 + 2x_2 + 3x_3 &= -1
\end{aligned}
$$

beginning with the initial approximation $x_1 = 1, x_2 = x_3 = 0$.

(2) Apply the Jacobi method and the Gauss–Seidel method to the system

$$
\begin{aligned}
2x_1 + \quad x_2 + \quad x_3 &= 7 \\
x_1 + 2x_2 + \quad x_3 &= 8 \\
x_1 + \quad x_2 + 2x_3 &= 9
\end{aligned}
$$

beginning (*a*) with the values (2, 3, 4), (*b*) with the values (1, 0, 0).

(3) Solve by an iterative method the system

$$
\begin{aligned}
3x_1 - \quad x_2 \quad\quad\quad\quad\quad &= \quad 2 \\
x_1 - 3x_2 + \quad x_3 \quad\quad\quad &= -1 \\
x_2 - 4x_3 + \quad x_4 &= -2 \\
x_1 \quad\quad\quad - \quad x_3 + 2x_4 &= \quad 2
\end{aligned}
$$

(4) Apply the Jacobi method and the Gauss–Seidel method to the system

$$
\begin{aligned}
x_1 + 2x_2 - 2x_3 &= 1 \\
x_1 + \quad x_2 + \quad x_3 &= 3 \\
2x_1 + 2x_2 + \quad x_3 &= 5
\end{aligned}
$$

Obtain the iteration matrices for the two methods and find their eigenvalues.

(5) Solve by iteration the system

$$
\begin{aligned}
5x_1 + 4x_2 + 3x_3 &= 12 \\
4x_1 + 7x_2 + 4x_3 &= 15 \\
3x_1 + 4x_2 + 4x_3 &= 11
\end{aligned}
$$

(6) Solve by iteration the system

$$8x_1 - x_2 + x_3 + x_4 = 4$$
$$x_1 + 5x_2 + x_3 - 2x_4 = 4$$
$$x_1 - x_2 + 10x_3 - x_4 = 4$$
$$2x_1 + x_2 - x_3 + 5x_4 = 4$$

giving the result correct to three decimal places. Find the solution also by a direct method, and compare the work involved in the two methods.

(7) By expressing the 4×4 matrix $\mathbf{A}$ in the form $\mathbf{LU}$, where $\mathbf{L}$ is a lower triangular matrix and $\mathbf{U}$ is an upper triangular matrix with unit elements in the diagonal, obtain the scheme for Crout's method for the solution of the system $\mathbf{Ax} = \mathbf{b}$.

Suggestions for further reading

Buckingham, R. A., *Numerical Methods*, Pitman (1957).

Bull, G., *Computational Methods and Algol*, Harrap (1966).

Collatz, L., *Numerical Treatment of Differential Equations*, Springer (1960).

Crandall, S. H., *Engineering Analysis*, McGraw-Hill (1956).

Fox, L., *An Introduction to Numerical Linear Algebra*, Clarendon Press (1964).

Hamming, R., *Numerical Methods for Scientists and Engineers*, McGraw-Hill (1962).

Hartree, D. R., *Numerical Analysis*, Oxford University Press (1957).

Hildebrand, F. B., *Introduction to Numerical Analysis*, McGraw-Hill (1956).

Kopal, Z., *Numerical Analysis*, Wiley (1955).

Lance, G. N., *Numerical Methods for High Speed Computers*, Iliffe (1960).

Lanczos, C., *Applied Analysis*, Pitman (1957).

National Physical Laboratory, *Modern Computing Methods*, H.M.S.O. (1961).

Ralston, A., *A First Course in Numerical Analysis*, McGraw-Hill (1965).

Redish, K., *Computational Methods*, English Universities Press (1961).

Index

THIS BOOK HAS BEEN SET IN MONOPHOTO TIMES NEW ROMAN
AND PRINTED AND BOUND IN GREAT BRITAIN BY
WILLIAM CLOWES AND SONS, LIMITED, LONDON AND BECCLES